Limnology and Ecological Preservation

LIMNOLOGY AND ECOLOGICAL PRESERVATION

Vierah Hulley

Delve PUBLISHING

www.delvepublishing.com

Limnology and Ecological Preservation

Vierah Hulley

Delve Publishing
224 Shoreacres Road
Burlington, ON L7L 2H2
Canada
www.delvepublishing.com

Email: orders@arclereducation.com

ISBN: 978-1-77469-452-7 (Hardcover)

ABOUT THE AUTHOR

Vierah Hulley (PhD) is an international expert in Environmental and Earth Sciences, with an extensive vocational background in the areas of environmental policy development, environmental and social risk management, natural resources management, sustainability, and climate risks management. She is the Founder of HL Nexus: Sustainable Management Solutions; a company dedicated to harnessing the power of data science in managing sustainability and climate risks. Vierah holds a master's degree in Geology and a PhD in Hydrogeology, with a focus on environmental and spatial science.

TABLE OF CONTENTS

LIST OF FIGURES

ABSTRACT

Limnology involves the integrated, multidisciplinary scientific research of inland waters –reservoirs, lakes, waterways, wetlands, rivers, and groundwater - as aquatic ecosystems interconnecting with their drainage basins and the atmosphere. Limnology involves the study of both freshwater and inland saline, or brackish waters, estimated to comprise approximately 45% of inland surface waters.

This volume advances the understanding of the bio-geophysical components of inland aquatic ecosystems of the drainage basin, movements of water through the drainage basins (running waters), and biogeochemical changes that occur on its way, and within standing (lentic) waters.Particular emphasis is placed on observing and analyzing approaches to ensure the health of these systems. In order to achieve better forecasts of climatic and ecological impacts on water systems and functions, conservation models must incorporate limnological tools.

Water plays a central part in the life on the planet. In our use and exploitation of water, human beings have had significant anthropogenic induced impacts on aquatic ecosystems. They manifest as stressors to the systems, such as in degraded water quality or loss of biodiversity. Given the significance of inland water systems to our wellbeing, in particular freshwater systems, limnology is a critical science to understand the mechanisms of natural systems. Increased understanding allows for more accurate predictive analysis of the same system under human-induced stressors, and the formation of a toolbox to proactively avoid or manage the changes.

PREFACE

Limnology is the scientific analysis of the interaction of chemical, physical, geological and biological attributesof all inland water bodies, within the continental borders. It has been a predominant sub-discipline of ecology since the early twentieth century. Besides, water bodies are typically encircled by large populaces and communities that depend on them for sustenance. For instance, freshwater lakes, a natural resource used for activities such as fishing, agriculture, and domestic consumption, this makes limnology a regular and practical science.

This volume elaborates on the science of limnology as well as the importance of its study, due to the continued global threats and degradation of freshwater systems. These significant sources of disturbance affect the biodiversity of fresh water bodies in variousways. The topics covered in this volume include; Chapter 1: Limnology and its Importance, Chapter 2: Formation, Zonation, Classification and Stratification of Lakes, Chapter 3: History of Limnology, Chapter 4: Limnology, Chapter 5: Polar Limnology, Chapter 6: Food Web Dynamics In Inland Water Systems, Chapter 7: Plants That Live On, In, Or Near Inland Waters, Chapter 8: Inland Water Resources Exploitation (Fishing, Mining, Power Generation, Transportation), Chapter 9: Global Crisis in the Freshwater Resources and Chapter 10: Aquatic Ecosystems Conservation and Management.

Contaminants such as toxic chemicals (heavy metals, chemicals and organic compounds) enter the aquatic ecosystem. While some are inactivated through chemical precipitation and oxidation, others have long term durations and persist within the natural system. Notwithstanding dilution and scattering in the aquatic ecosystem, biomagnification of metallic and organic elements is typical, which significantly raises their toxicity. The ecological impacts of some of these elements remain unclear. This eutrophication rate prompts higher levels of decay. Extreme loading of organic material emerges from organic wastewater from, industrial effluents, human settlements and excessive nutrient spillover from agricultural activities.

Another component that contributes to decline in aquatic ecosystems is the suspension of fine organic and inertinorganic particles. The breakdown and transport of these residues are high as a lot of vegetation between water and land is being removed, essentially for agricultural activities.

Some authors argue that the inclusion of some species can bring about risks to biodiversity. Speeding up human-based changes in hydrological systems related to flood management and water supply changes further prevent effective usage practices.

Management and prevention of degradation needs resourceful economic appraisal of freshwaters for proficient use of water supplies for horticultural, industrial and domestic purposes. Changing resistance, steadiness and biodiversity of freshwater environments: a focal part of ecological stability is the capacity of an environment to get back to a consistent equilibrium after a disturbance.According to the volume, decreased biodiversity improves vulnerability by lessening the aggregate physiological resilience of the community to massive habitat changes/unsettling conditions. Differences in plant productivity can cause different forms of organic matter chemical changes. Due to large amounts of particulate from plants, heterotrophic uses of organic matter may be slower.

Adjustments to the quality and amount of disintegrated organic content entering the aquatic environment from the drainage basins will impact this intrinsically chemical-based metabolic balance. In this circumstance, undisturbed riparian floodplains, forested watersheds and wetland/littoral regions are significant to the metabolic balance inside water ecosystems.

CHAPTER 1

LIMNOLOGY AND ITS IMPORTANCE

CONTENTS

1.1 INTRODUCTION

Figure 1.1: Limnology is the study of biological, physical, chemical and biological attributes of inland waters such as the lake pictured.

Source:https://media.istockphoto.com/photos/view-of-lake-bohinj-in-slovenia-picture-id1247151211?k=20&m=1247151211&s=612x612&w=0&h=DA7_lahxcK80eYFGr0k9nr8z2jEGucdjL38ET0MOdts=

Limnology is a field of study that encompasses a broad variety of topics related to inland waters. More significant than any formal explanation is the fact that limnology investigates inland rivers as natural processes. As a result, limnology may be regarded a field that includes information from all the other disciplines that contribute to the study of inland waters (Wu et al., 2021). This is because limnology focuses on the study of freshwater ecosystems. The study conducted in the discipline of limnology has an impact not only on the science of fisheries but also on hydrology, oceanography, and several subfields within geology, botany, biology, and environmental engineering. It is not unheard of for notable limnologists and specialists from other fields to pen papers in which they express their worries about the field's current status and prospects, as well as the pressing need for change. From this vantage point, several topics, such as the teaching of limnology in educational institutions, the availability of funds for fundamental limnological research, and the acknowledgment of limnology by other academic domains, have gained prominence.

Within the academic field of limnology, there are a few distinguishing characteristics that point to intellectual activity as well as an appropriate concentration on conceptual and societally important issues. Researchers are looking into the nitrogen cycle, with a particular emphasis on how aquatic ecosystems react when nitrogen is added to their environments. Inland waterway microbiological activities, the origin and transformation of organic matter in surface waters, the impact of ultraviolet radiation on aquatic biota and aquatic processes, and the principles that guide the construction of aquatic communities are some of the other topics that will be covered in this volume. Not only is the quantity of limnological study articles growing in number, but so too is the growth of scientific organizations and the number of people who belong to them. The International Society of Limnology (ISL) concluded that there are several subfields of limnology that have room for advancement but currently do not (Tundisi et al., 2010). Only a few examples of these areas of concentration are provided here, but some of them include the ecosystem view, the sub-disciplinary balance, and communication with other fields of study, application ties, and research support. This continues to be the case even though contemporary limnology provides a great deal of cause for optimism.

Figure 1.2: Inland water systems are rich ecosystems that support different flora and fauna.

Source: https://shuswappassion.ca/wp-content/uploads/2014/03/limnology.jpg

Obtaining a doctoral degree, which is required for employment in academic settings, is presently the primary purpose of limnology education. Even though some graduates do find employment in the public or private sector, the majority of large educational institutions place a high priority on preparing students for academic careers in places like colleges and universities. This traditional emphasis appears to be irrelevant given the increased demand for limnological competence in settings other than academic institutions, which calls into question the relevance of the traditional focus. This change in emphasis may call for significant adjustments to be made to the manner that limnology is taught (Togaki & Katano, 2020). When it comes to the study of limnology, a high level of specialization in a specific sub-discipline or topic is not typically related to the production of individuals who will engage in activities outside the confines of the academic sphere. Education at the postgraduate level in the field of limnology may need to be widened and standardized.

This might be accomplished by the implementation of a certification program or a consensus-based standard that ensures both the breadth and consistency of education among postgraduate students. Limnology is a discipline that falls under the umbrella of ecological science. Limnology has evolved into a more specialized field over the past several decades, with sub-disciplines centering their attention on certain ecological components. This study is significant when considered in the context of broad frameworks; yet, because there has been no attempt at system integration, the value that can be derived from this research has been severely constrained. Limnologists are obligated to develop a solution to this conundrum to fulfill their research obligations. Because limnology deals with concerns that are biological, physical, geological, and chemical in nature, there has to be sufficient progress achieved in some subfields of research to allow others to continue expanding. In the study of limnology, there is a significant disparity in the proportions of some of the species. Chemical limnology is seen as being of more importance than physical limnology in the context of North American research (Teffera et al., 2019). Research on plants and microbiology has been pushed to the background in favor of studies on the photosynthesis and decomposition of animals. They endanger an industry that gets the vast bulk of its power from working together with other people. Some of the most fruitful areas of collaboration with limnologists are in fields that have nothing to do with the study of limnology.

In the present day, some ecologists believe that limnology has no connection to the study of ecology. Hydrology and limnology, for instance, may be at a fork in the path at this particular juncture in time. The connection between limnology and oceanography appears to be strong, notwithstanding the concerns of certain oceanographers who are concerned about the possibility of the extinction of limnology. The field of limnology in the United States does not have a significant relationship to real-world applications. Evaluations of water systems are frequently carried out without the assistance of limnologists, even though it is abundantly clear that such knowledge is required. This is puzzling in light of the fact that limnology places a high premium on questions concerning anthropogenic origins (eutrophication, acidification). There may be weak linkages as a result of older social attitudes that prioritized the extraction and exploitation of particular resources over concerns for the long-term health of the ecosystem and multifunctional management. There is a chance that the destruction of ecosystems was caused in part by people holding these perspectives (Alcocer et al., 2021). The old model is being replaced by a new one, as demonstrated by the recent embrace of ecosystem management by governments. Even if limnology is already very well equipped for this paradigm shift, the discipline still has to completely express itself in order to be able to participate in the shift. An increase in funding for research might be beneficial to a wide variety of areas, including limnology.

Despite this, the Challenges Committee arrived at the conclusion that the financing for limnology research is insufficient in comparison to the need that society has for knowledge of inland waterways. On the other hand, new developments at the EPA could be able to help improve the widespread lack of postdoctoral training opportunities that have been a problem in the past. There are no specialized educational programs or activities sponsored by the government that are geared toward the expansion of limnology. The only exception to this is limnology focused on the Great Lakes. To research interconnected systems, educational institutions frequently lack the necessary infrastructure, which is extremely expensive. Both the national efforts to conduct research and the education of graduate students have suffered as a result of university research programs falling behind government research facilities. This situation has been a source of contention. Limnology has the potential to develop further and become more established if scientific societies are actively involved (Soranno et al., 2010). Since its inception, the primary focus of the American Society of Limnology and Oceanographic (ASLO), which is recognized as one of the most important limnological associations in North America, has been almost exclusively on the dissemination of scientific accomplishments in the field of limnology and oceanography.

Figure 1.3: A scientist measuring the health of an inland water body.

Source:https://lgl.com/images/headers/oceanography-bg.jpg

This is not the case, even though limnology and oceanography are at the forefront of contemporary scientific findings. The mix of limnology and oceanography that is included in ASLO is beneficial from an academic standpoint, but it may have made it more challenging for limnologists to link their interests through ASLO. There are important subfields of limnology that do not now satisfy the professional requirements set out by the American Society of Limnology, and it is conceivable that this pattern will persist in the not-too-distant future (Sigala et al., 2017).

1.2 IMPORTANCE OF LIMNOLOGY

The basic objective of limnology, like that of other scientific subfields, is to get to the bottom of things and figure out what is going on. The following are some rules of thumb that can be applied in a variety of decision-making settings (Simões et al., 2011).

Because of its ability to make predictions, which are crucial in the realm of applied limnology, the study of limnology is recognized as a scientific discipline. Acid rain has been created in recent years as a result of coal-burning power plants, factories, and automobiles. This has led to the progressive deterioration of inland water bodies. This has had a significant influence on the rate at which the climate is changing. Inland water bodies can be returned to their former state by putting a stop to all of these potentially damaging activities and even reversing some of them.

What is Acid Rain?

This is comprised of profoundly acidic water beads because of air emissions, mostly sulfur and nitrogen from vehicles and assembling processes. It is much of the time called acid rain as it contains many kinds of acidic precipitation.

It happens in two ways: wet and dry. The Wet type is any type of precipitation that eliminates acids from the air and puts them on the outer layer of the earth. Without a trace of precipitation, dry showers of particles and gases adhere to the ground through residue and smoke

Reasons for Acid Rain

Acid rain comprises of acidic droplets which get blended in with the fresh rain water. Sulfur, carbon dioxide and nitrogen particles that are infused with water are formed in two ways either man-made i.e as the emissions that are given out from transport or by regular causes like lightning in the environment delivering nitrogen oxides and vehicle emissions causing carbon dioxide and sulfur oxides. As per the Royal Society of Chemistry, the phrase acid rain was established in 1852 by Scottish physicist Robert Angus Smith. He settled on the word while studying water science close to modern towns in England and Scotland. The typical downpour we experience isn't spotless clean i.e. water and carbon dioxide typically combine together to create acid rain which can be quite destructive.

The pH rating of normal rainwater is around 5.7, which is quite acidic. The oxides of sulfur and nitrogen are dissipated by the wind alongside the dust particles. They settle on the earth's surface subsequent to descending as precipitation. Acid rain is basically a side-effect of human activities which discharge nitrogen and sulfur oxides into the environment. For instance - the burning of fossil fuels and, unethical waste discharge removal processes. Nitrogen dioxide and sulfur dioxide go through oxidation, and afterward, they combine with water bringing about the development of sulfuric and nitric acids.

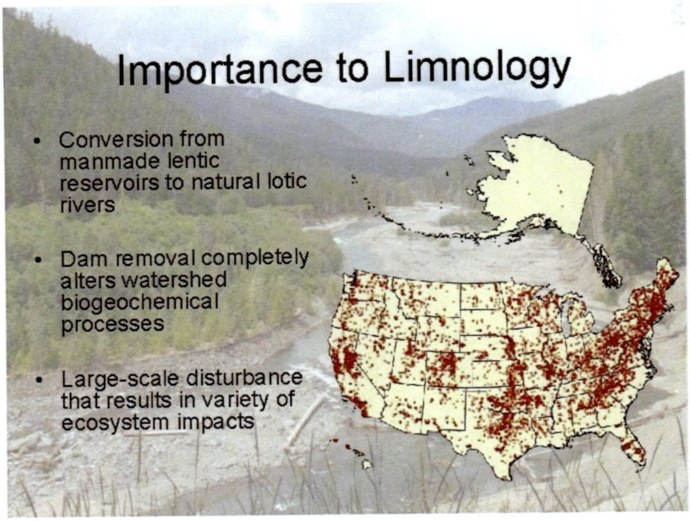

Figure 1.5: Importance of Limnology.

Source:https://slidetodoc.com/presentation_image_h2/7bf64786b1b551ed447 3dd807b6b632c/image-5.jpg

When people interact with the natural world including aquatic flora and fauna, they are responsible for a wide range of structural changes in the water (overexploitation of aquatic plants and animals, introduction of foreign species). Inland water basins have the challenge of eutrophication in addition to the problem of pollution.

We need to do a better job of managing our lakes, rivers, and reservoirs if we want to maximize their potential. The construction of aesthetic dams has had a considerable negative influence on the natural aquatic and terrestrial ecosystems of several different countries. Because of these developments, new ecologies have developed, and each one has its unique set of traits that set it apart from the others (Abubakar & Abubakar, 2013). By preserving these ecosystems with the use of fundamental scientific knowledge, it will be possible to do so at a low cost and for a broad variety of purposes. In addition to this, there is a possibility that it will pique the interest of limnologists and further expand their conceptual understanding.

The study of lake and reservoir development is an additional component that differentiates basic limnology from applied limnology as a subfield. To figure out the ecosystem, it is required to evaluate the origin of lakes and reservoirs and determine comparative criteria. When land and sea ecosystems are brought together, it is feasible to produce more diverse and intricate biological communities.

In recent years, as freshwater fisheries, in particular fisheries biology, has emerged as a specialization with limnology. Throughout the twentieth century, important research traditions in limnology became increasingly focused and externalized, which ultimately led to a progressive split between the marine and inland systems. The science of fisheries is being reevaluated because of a paradigm shift in freshwater fisheries research, which is leading to a redefinition of the objectives and methodology of research.

Research on fisheries must adopt a multimodal, multidisciplinary, and even transdisciplinary approach if we are ever going to comprehend them as socioecological systems. The purpose of this study is to investigate fisheries from both a social and a natural point of view, and to do so, it integrates the most beneficial aspects of both worlds. Even though some aspects of social-ecological fisheries systems are still important to understand, limnology has evolved into a sub-discipline of fisheries study rather than the fundamental supporting science it was formerly supposed to be. This is even though there are still some aspects of social-ecological fisheries systems that are important to understand (Barbosa et al., 2012). Even if the study of limnology is no longer considered important, this does not imply that it is no longer required. It will be necessary for limnologists and scientists who study freshwater fisheries to undertake substantial paradigm shifts for them to improve their academic interactions.

The importance of water for irrigation, household and industrial use, generation of hydroelectricity, waste disposal, recreation, and the support of aquatic life, including commercial and recreational fishing, has led to a conflict between the conservation and utilization of aquatic resources. This conflict has arisen because water is required for these activities. This problem is caused by the fact that water is such an important resource. For a complete analysis of this issue from a scientific point of view, it is important to do research in the field of limnology, which focuses on the study of aquatic environments. To properly conduct this study, aquatic habitats need to be well researched.

As part of a comprehensive and far-reaching project to maintain water quality in the United States, all pollution caused by rivers in the United States will be eradicated. This will be done as part of the program. This devotion has been symbolized by the Clean Water Act of 1972 and its subsequent amendments. This dedication has gradually been translated into law, which has resulted in the expenditure of billions of dollars to protect aquatic regions (about 600 billion dollars since 1972). Additionally, ambitious physical protection alternatives are being developed, particularly for wetland areas.

Figure 1.6: Fisheries Acts protect fish stocks from depletion caused by overfishing.

*Source:https://www.wcel.org/sites/default/files/styles/page_image/public/2016_enviroreviews_button_fisheries.jpg?itok=Ww_bM05_*

The Fisheries Act of Canada (2020) includes provisions for the establishment of stringent restrictions, which are intended to protect the aquatic ecosystems of the nation. The only way for society to make the most of its ever-increasing protection budget is for researchers to speed up the rate at which they collect data on the functioning of aquatic ecosystems. It is necessary to have a complete awareness of the different stressors that have the potential to affect aquatic ecosystems, as well as a deep comprehension of how the stress response might alter depending on the geological or climatic elements that are present. These details are required to be included in regulatory systems (Balarabe, 2006).

At the same time, concerns must be addressed regarding the capacity of the public sector to both recover and remain resilient in the face of stress. The environment and the need from the public to address certain environmental challenges, such as the acidity of surface water, pollution, physical change, and harm to aquatic ecosystems, should have led to a rise in the amount of research conducted on aquatic ecosystems. One of these problems is the acidification of surface water. Others include the contamination of organic waste and the destruction of aquatic ecosystems. On the other hand, the rate of limnological research has not been much impacted by the significant rise in public financing for the conservation of aquatic environments. In the past, government support for research has been very minimal. In the past, these contributions have either been rapid and superficial practical studies or a synthesis of the current literature on subjects that are crucial to the creation of new laws. Since the implicit expectation that limnology will continue to supply a steady stream of critical knowledge with few extra resources is expected but not clearly articulated, new legislation and related expenditures are necessary. Aside from the fact that it's implausible, it's also impractical. Without a strong tie to research, the government's resources could be squandered if protection systems continue to develop without a strong link to research.

The potential of the study of limnology has been brought to light by two key efforts that have been undertaken in the past two decades. An extensive study effort with a well-defined scientific goal was carried out at the Canadian Experimental Lakes Area (CELA). In the spirit of scientific experimentation, several lakes were altered in a region that had both manipulated lakes and unaltered lakes, so that a comparison could be made between the two types. The Hubbard Brook Experimental Forest is a good illustration of the potential of limnology since it was utilized to tackle several problems associated with watershed management by modifying several smaller watersheds in a laboratory setting. The study of limnological ecosystems frequently makes use of comparative and experimental approaches; because of these two enormous initiatives, the feasibility and adaptability of these methodologies were put to the test (Shah et al., 2019). Limnology is currently an essential component in the conservation of animal habitats as well as in the management and administration of water resources. Limnologists are involved in a wide variety of activities, including the management of lakes and reservoirs, the avoidance of water pollution, the preservation of streams and rivers, the development of artificial wetlands, and the enhancement of wildlife populations.

1.2.1 The Importance of Limnology to Society

Water plays a critical role in agricultural and domestic consumption; hydropower generation; waste disposal; recreational activities; and aquatic life, including commercial and recreational fishing. This has resulted in a conflict between conservation and use of aquatic resources. This topic necessitates an understanding of aquatic ecosystems, which are extensively studied through limnology. Aquatic Ecosystems Must Be Safeguarded -this is the most ambitious water quality protection program the United States has ever devised, to eliminate all pollution-related consequences in American rivers. This commitment has been gradually implemented from law to regulation since the Clean Water Act of 1972 and its subsequent amendments, necessitating the investment of several billion dollars to protect aquatic environments (cumulatively around 600 billion dollars since 1972: U.S. EPA 1991). Environmental Protection Agency of the United States, 1991: (about 600 billion dollars in total since 1972). Comprehensive physical safeguards, particularly for wetlands, have already been established, by the National Academy of Sciences (Reed et al., 2008).

The Fisheries Act governs aquatic habitat protection in Canada, which follows a similar set of rules. Only by rapidly acquiring knowledge about the operation of aquatic ecosystems can a society's protection expenditures be maximized. Aquatic ecosystems are extremely sensitive to various types of stressors, and regulatory systems must consider this when deciding how to respond to stress under different geological or climatic conditions. Understanding stress recovery and recovery mechanisms must also influence the distribution of public funds. An increase in aquatic ecosystem research should have been justified in light of environmental legislation and public pressure to address specific environmental issues such as acidification of surface waters, metal and organic compound pollution of waters, and physical alteration and impairment of aquatic environments. Despite significant increases in public funding for aquatic habitat conservation, limnological research has slowed down.

Governments have frequently given little thought to research, typically through short and shallow applied studies or the synthesis of existing literature on topics critical to the creation of new legislation. New regulations and associated expenditures have been developed with the implicit but unrealistic assumption that limnology will continue to provide a steady stream of vital information with minimal additional resources. They are increasingly at risk of losing focus and wasting money if they do not have a strong link to research in the development of new protection systems.

A Look at Limnology's Future: Two major limnological experiments at the end of the twentieth century demonstrated what could be done with this new field of study. An aggressive, experimentally-oriented program involving the manipulation of numerous lakes in a region where manipulated and manipulated lakes could be compared in the spirit of experimental science was brought to a logical conclusion in the quest to understand aquatic ecosystem function in Canada's Experimental Lakes Several studies have discovered that (Several studies have discovered that (Pinilla & Gabriel, 2016). The experimental manipulation of small watersheds at Hubbard Brook Experimental Forest provided clear and fast answers to hundreds of underlying concerns about management and use of the potential of limnology in a similar manner and Likens. These two extensive studies highlight the comparative and experimental approaches that are unique to limnology ecosystem study.

Figure 1.7: Inland water bodies help with the regulation of carbon dioxide.

Source: https://image.shutterstock.com/image-vector/limnological-disaster-emission-carbon-dioxide-260nw-1984559921.jpg

1.2.2 Importance of Limnology as a Science

The search for fundamental truths is central to limnology, as it is to other fields of study. Management and decision-making principles are applicable in a variety of situations. The comparative aspect of limnology should be highlighted especially. When compared to the hydrology of river systems, lakes, and reservoirs, many practical factors influence aquatic organisms' life cycles and distribution. As an example, it is also necessary to consider phytoplankton physiology and how it responds to changes in light intensity caused by currents.

According to this new approach to the system, hydrodynamics and their effects on vertical structure are a major controlling factor in the phytoplanktonic succession of the system. As a result of this method, numerous theoretical and practical avenues for limnology research have been opened up. Unlike oceanography, modern limnology has a strong conceptual similarity to it. Limnology is heavily used in applied limnology because it can predict the future. In recent years, acid rain has increased due to deforestation and air pollution near waste reservoirs. All of these processes that harm inland water bodies can be altered and corrected (Pinilla, 2010). Human interference with aquatic ecosystems, on the other hand, has resulted in numerous structural changes (overexploitation of aquatic plants and animals, introduction of exotic species). Eutrophication and pollution both have an impact on inland waterways.

To maximize the potential of lakes, rivers, and reservoirs, proper management is required. Landscape dams are not uncommon in changing the natural ecosystems of many countries, resulting in the formation of new ecosystems with distinct characteristics. This knowledge can be used to manage these ecosystems for a variety of purposes with little investment. Limnology can benefit from this as well, by encouraging scientific curiosity and broadening fundamental knowledge. The focus of applied limnology on the evolution of lakes and artificial reservoirs distinguishes it from basic limnology. Scientists can identify the ecosystem by comparing lakes and reservoirs from their origins.

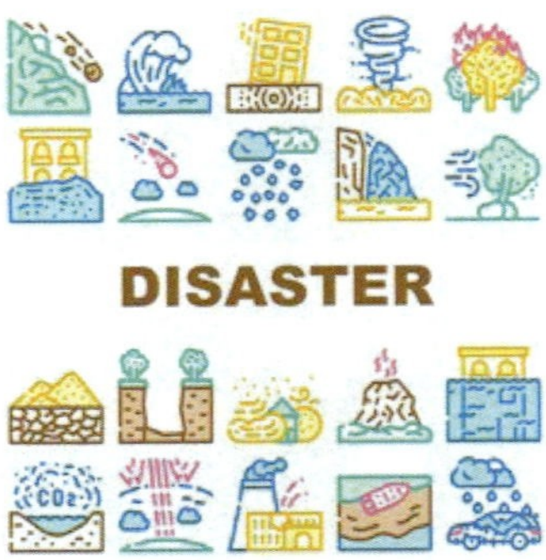

Figure 1.8: Disasters like flooding and volcanic eruption.

Source: https://image.shutterstock.com/image-vector/disaster-destruction-collection-icons-set-260nw-1979988467.jpg

CHAPTER 2

FORMATION, ZONATION, CLASSIFICATION AND STRATIFICATION OF LAKES

CONTENTS

2.1 FORMATION

One must understand where a lake originated and how it was formed to comprehend its structural properties. The present physical, chemical, and biological state of a lake are determined by several variables, including the lake's genesis, the size and form of its basin, the terrain and chemistry of its watershed, and the climate of the surrounding region, and the presence of local biological species. The conduct of humans has changed during the past century. The United States of America is home to a dizzying number of freshwater lakes. The glaciated landscapes of the Great Lakes, limestone sinkholes in Florida, and alpine lakes in the Pacific Northwest; all of these may be found in the United States. The Lake Ecology area of Water on the Web focuses on the lakes that are found in the glaciated landscapes of the Great Lakes region (Simmatis et al., 2021). There are 12,034 lakes in the state of Minnesota that are greater than 4 hectares. The bulk of Minnesota's lakes were formed by glaciers that blanketed much of the state 12,000 years ago and are still there now.

As time passed, the ice blocks that glaciers had buried beneath the soil melted, resulting in the excavation of holes and the dumping of debris across stream banks. These naturally occurring depressions or impoundments eventually became lakes after they were flooded with water.

As the glaciers retreated, a layer of silt was left behind in the deepest areas of the lake. These sediments, which reached the lakes from the rivers, were responsible for the decomposition of the organic compounds that were present in the watershed, as well as the aquatic plants and algae. In a normal lake in Minnesota, one can find 9–12 meters of material with similar characteristics at the lake's bottom. Sediment deposits on the bottom of a lake help to ensure that the history of the lake is not lost. Paleolimnology is the scientific name for the study of lake sediments.

Paleolimnologists gather sediments from lakes so that they may investigate the physicochemical and biological changes that have occurred in lake environments over time. Lead-210 and carbon-14 are the radioisotopes that are utilized the most frequently in the process of dating lake sediments. The age of the sediment sample may be determined based on the rate of radioactive decay of an isotope. The presence of a large quantity of pollen from ragweed and other plants in the core of an artefact can be interpreted as evidence of agricultural soil disturbances or deforestation by several alternative dating techniques. Investigations of stratigraphic sediment are being conducted increasingly often to learn more about people's impact in lakes. The number of diatoms and plant pigments present in a lake may be used to assess the pH of the water, how clear the water is, and how productive the algae are (Robarts & Zohary, 2018). Analysis of the inorganic and organic components of the soil, in addition to testing for metals, can be used to make predictions about the rate of soil erosion. In recent years, many analytical methods, including sediment dating, sediment coring, and others, have been utilized in the state of Minnesota to investigate patterns in mercury levels found in lake sediments across both space and time. The use of this data allowed for the estimation of mercury trends as well as state-based sources.

People tend to picture lakes as being similar to bathtubs, with the water being distributed uniformly from the top to the bottom, sides to sides, and front to the back of the body of water. Lakes may be discovered in a wide variety of geographical environments. Lakes may be described as having a diverse variety of properties when viewed in terms of their physical, chemical, and biological make-ups. The quantity of light that penetrates a lake, the temperature of the water, and the flow rate of the water are all examples of the diverse physical qualities that lakes possess.

Figure 2.1: A mountain lake.

Source: https://images.pexels.com/photos/11967510/pexels-photo-11967510.jpeg?auto=compress&cs=tinysrgb&h=750&w=1260

Lakes are distinct from one another in terms of the primary ions, pollutants, and nutrients they contain. It is essential to understand the structural and functional distinctions that exist between lakes due to the large variety of bio-physiochemical characteristics that separate them from one another. All these variables exhibit significant geographical and temporal fluctuation on several timescales, including minutes, hours, and days, seasons, decades, and geological epochs. Similar to the way that a forest is organized from the ground: up from the canopy to the roots, lakes are composed of numerous physically distinct layers (Ribas et al., 2019).

The relationships between light, temperature, and wind are perhaps the most fundamental aspect of lakes. The amount of light that is absorbed and diffused by the water column has a significant impact on the temperature as well as the photosynthetic potential. The process of photosynthesis provides sustenance for a major proportion of the food web. In addition to this, a significant quantity of oxygen is discharged into the ocean. Because of solar radiation, wind patterns and water circulation both affect the flow patterns that occur inside the lake basin. As you descend deeper into the water, the light intensity at the lake's surface changes depending on the season and the amount of cloud cover, and it gets dimmer as you get closer to the bottom. When there is more light available for the water column to take in, there is a greater potential for photosynthesis to take place. Some algae float in water, some algae attach themselves to surfaces, and aquatic plants with vascular systems all fall under the category of photosynthetic organisms.

The rate at which light weakens as it penetrates deeper into the ocean is determined by the amount of light that is absorbed and scattered by dissolved chemicals. These molecules both absorb and scatter light. The vertical extinction coefficient is defined as the percentage of light at the surface that is either absorbed or scattered by a vertical column of water that is one meter in height. The letter k is used to indicate this particular parameter. Lakes with low k-values allow more light to pass through them than lakes with high k-values do because of this difference.

The greatest depth at which algae and macrophytes may develop is directly proportional to the amount of light that is available. Limnologists believe that a lake has reached this depth when the amount of light that can reach the bottom of the lake is between 0.5 and 1 percent of the amount of light that can reach the surface of the lake. The name for this region is called the euphotic zone. According to estimations derived from the Secchi disk, the visibility limit occurs at a distance that is two to three times greater than this depth (Reed et al., 2008). Light may be measured in many different ways due to the many different attributes it possesses. The author urges the reader to pursue further knowledge by reading additional books. Variations in the amount of light that can penetrate a lake can have several direct and indirect effects on the chemical and biological processes that take place there. Human activity, in conjunction with other land-use activities that occur within the watershed, is the most prevalent cause of noticeable shifts in the transparency of lakes.

Figure 2.2. An Oxbow lake.

Source: By Oliver Kurmis, CC BY-SA 2.0 de, https://commons.wikimedia.org/w/index.php?curid=241148

In regions with a temperate climate, the depth of the water column is extremely cold and practically constant in temperature after the ice has melted. In the spring, as the brilliant sunshine is absorbed, the water column begins to warm up, which in turn causes the average daily air temperature to rise. If there is no wind, a plot of surface temperatures vs. subterranean temperatures may look somewhat like this, with depth rapidly decreasing: This pattern is susceptible to being severely disrupted by the density of the water, which is an additional physical aspect. Only water's density is lower in its liquid state compared to its solid-state. As a direct consequence of this, ice floats in the water whereas water with a temperature just above freezing, sinks. When the bulk of the components in a material change their condition from liquid to solid, the resulting substance will be denser. When heated over 4 degrees Celsius, the air has the greatest density; however, water begins to lose density.

2.2 CHARACTERISTICS OF LAKES

I) Lakes are inland depressions containing water.

These massive basins are created in different ways, for example, ice engravings developed through moving ice sheets, deserted pieces of waterways, volcanic pits, and more. Be that as it may, the most well-known way is through changes in plate tectonics. When the Earth's crust moves and makes faults, they form land depressions that are filled with water after some time through precipitation or flowing streams. Most lakes have outlets underneath or over the water surface, where water is lost through the surface or underground. Then again, closed lakes mostly lose water via evaporation.

II) Lakes are still, sluggish moving waterways.

Lakes are defined by standing, sluggish waters. Perhaps the most straightforward method for recognizing a lake is through its water cycle. Lakes are viewed as lentic environments portrayed by as yet, standing, or sluggish waters. Waterways and streams stream consistently from a water source down to the stream's mouth interfacing it with the ocean. Not at all like streams, lakes are quiet and don't show waves or disturbances. Lakes likewise appear to be unmoving on a windless day and possibly show developments whenever exposed to external factors like winds. Then again, waterways will keep flowing even without the impact of wind.

III) Lakes are homes to complex niche environments

Lakes are complex environments. Some aquatic life can only make due in freshwater. One novel quality of a lake is its capacity to sustain complex biological systems. Environmental systems differ as every creature species and plant is different. Lakes encourage complex biological systems like animals and plants living in their natural surroundings and with changing necessities and ecological needs. The type of fish and other marine biodiversity living in a waterway can likewise determine whether it is a lake or not. The trout, for example, is a cold freshwater fish that has developed in deep lake biological systems.

2.3 SEASONAL STRATIFICATION OF LAKES

2.3.1 Spring

Before the spring ice cover melts, the water temperature at the lake bottom is generally 4 degrees Celsius in the upper Midwest and higher altitudes. The water above this layer, which will be near zero degrees Celsius, will have a substantial temperature differential from this layer, which has a temperature difference of only a few degrees. As the temperature rises, the ice begins to disintegrate and melt. When water at the surface is heated, its density begins to drop. It just takes a small amount of energy from the wind to properly mix a lake such that the water temperature at the top and bottom of the lake is the same. This is referred to as "turnover" in the industry. Even after winter is over, the water at the surface of the earth will continue to soak up the sun and warm up (Pinilla & Gabriel, 2016). When the temperature of the water rises, the upper layer of water becomes more transparent in comparison to the layer below it. It is conceivable for winds to stir up the lake from the bottom to the surface for an extended period; however, the water at the surface will become too buoyant and heated to properly stir up the water that is deeper and denser.

Figure 2.3: The dam is a manmade inland water reservoir.

Source: https://www.ejatlas.org/conflict/bc-hydro-site-c-dam-canada

2.3.2 Summer

The temperature differences between the top and lower water levels grow increasingly pronounced as the summer unfolds. In deep lakes, the three physical layers that are most commonly found are the epilimnion, the metalimnion, and the hypolimnion. The epilimnion is the name given to the uppermost layer, which is warmed and well mixed. The metalimnion, which is often referred to as the thermocline, is a layer of water that experiences a considerable temperature drop with increasing depth. The hypolimnion is formed when the metalimnion divides the hypolimnion and the epilimnion into their respective halves.

During the summer, there is a shift in the density of the metalimnion, which prevents the top and bottom layers from mixing and creates a physical barrier. The exposure to wind and the amount of water in a lake are the two parameters that have the most impact on the mixing depth. Likely, lakes ranging from small to medium in size and located in northern climatic zones are stratified and poorly mixed down to depths of between three and seven meters. During the summer months, the depth of large lakes can range anywhere from 10 to 15 meters (Pinilla, 2010).

2.3.3 Autumn

When the temperature drops in the fall, the density difference between the epilimnion and the hypolimnion decreases. The thermocline will go deeper into the sea as time goes on because the wind will agitate the water to higher depths. The fall winds, which cause the lake to "turn over," cause the temperature and density levels to become more balanced across the body of water. As the air temperature drops, the water at the surface of the lake begins to freeze. The density stratification that occurs beneath the ice is not as noticeable throughout the winter as it is during the warmer summer months.

As a consequence of the majority of the column having an isothermal temperature, the water above the ice has a density that is higher by 4 degrees Celsius than the water below the ice. Because of the little difference in density between water between 0 and 4 degrees Celsius, stratification is less likely to be maintained over time. On the other hand, the ice that is floating on top of the water shields the water column below it from the turbulence that is caused by the wind (Bassi et al., 2014). As a direct consequence of this, the concept of layering will be a recurrent one throughout the winter.

Figure 2.4: A cool autumn weather.

Source:https://images.pexels.com/photos/1477199/pexels-photo-1477199.jpeg ?auto=compress&cs=tinysrgb&w=1260&h=750&dpr=1

2.3.4 Winter Stratification

Lake thermal stratification refers to an adjustment of the temperature at various depths of the lake, caused by the adjustment of water density and temperature. Cold water is much denser compared to warm water and the epilimnion mostly comprises of water that isn't so dense as the water in the hypolimnion. Nevertheless, the temperature of greatest density for freshwater is 4 °C

During the fall season, cold air temperatures chill the lake's surface. As surface water chills, it becomes denser and sinks to the bottom. Ultimately, the whole lake comes to around 39EF (4EC). When the surface water cools further, it becomes less dense, "floating" above the denser 39EF water, causing ice at 32EF(0EC). The lake water beneath the ice remains around 39EF. This condition is known as winter stratification, the ice cover keeps wind from churning the water.

2.4 OVERVIEW

Stratification in shallow lakes occurs seldom, if at all, during the summer. During the summer, polymictic lakes see frequent stratification and destratification. It is possible to achieve complete reoxygenation of the water column in dimictic circumstances, which may be found at the West Upper Lake Station of Lake Minnetonka, Lake Independence, and in a great number of other lakes located in the Upper Midwest (Pérez-Fuentetaja et al., 2006). The lake may not be very large, but despite its size, it has a strong protective system and is rather deep. Because of the open-pit mining that occurs in the region, the lake depths in northern Minnesota are disproportionate to the size of the lakes. It is projected that these lake characteristics will completely mix in the fall before the lakes freeze over.

Despite the absence of evidence, it seems improbable that the water in some deeper mine pit lakes will mix to the bottom. Rarely do you come across lakes that have poor circulation and tend to build up a layer of stagnant bottom water? These bodies of water are referred to by the name meromictic lakes. Both stasis and anoxia in the subsurface water can persist for a considerable amount of time.

The majority of meromictic lakes have average depths that are rather deep, measuring between Because of the basic shape of these lakes, they are often on the smaller side and are protected from the wind. In this particular instance, density changes brought on by temperature have a less significant role than those brought on by the high concentration of dissolved components in the monimolimnion. Large lakes that rarely freeze over tend to be monomictic, meaning that they mix in the autumn, winter, and spring but stratify during the warmer summer months.

The term "watershed" refers to the land and water region that collects runoff and eventually empties into a particular river or lake. As a consequence of this, the lake that was selected will act as the reference point for creating a watershed. It is conceivable for watersheds to incorporate sub-watersheds within their boundaries. As an illustration, the watershed of a river that drains into a lake is a component of the lake's overall drainage basin. A lake serves as a mirror that reflects the surrounding watershed. A lake, for instance, can give information on the size and geography of the watershed, in addition to its geology, land use, fertility, and erodibility of the soil (Papa & Mamaril, 2011). The influence that the watershed has on the nutrient load may be seen through the connection between nutrient loading and the ratio of watershed surface area to lake surface area.

The more land that is covered by a lake's watershed, the poorer the lake's water quality will become. This becomes readily apparent as the catchment area of the lake expands, as does the number of runoff sources that feed into the lake. When the watershed is wider, there is a greater chance that precipitation-fed water will reach the soil and take up minerals than when the watershed is narrower. Seepage lakes are only found in extremely tiny watersheds that receive the majority of their water supply from the earth. Drainage lakes are lakes that receive the majority of their water supply from the rivers and streams that flow into them. Drainage lakes are often found in mountainous regions. Because of the connection between the watershed and the lake region, seepage lakes have higher quality water than drainage lakes. Seepage lakes are particularly susceptible to acid rain because they do not possess sufficient buffering capacity.

Both the quality and quantity of the water in lakes may be affected by land usage. The large concentration of impermeable surfaces in metropolitan areas prevents precipitation from seeping into the soil and increases the pace at which water from the surface flows into lakes and rivers. Cities with high flushing rates have the potential to move large particles from rivers and streams into lakes. As a consequence of this, the quantity of water might influence the quality of the water. Water picks up dissolved nutrients and contaminants as well as particle matter as it is flowing over pavement or parking lots.

Even in the absence of any living species, a lake will nevertheless contain a diverse variety of molecules and ions. This is because the weathering of soils in the watershed, the atmosphere, and the lake bottom all contribute to the composition of the lake. This indicates that temperature and the geology of a lake's basin are the two most significant elements that play a role in defining the chemistry of a lake. The lakes each have an equal concentration of three major anions and four important cations.

Ion balancing demonstrates, via the utilization of equivalents, that the number of cations with a positive charge is equivalent to the number of ions with a negative charge. In addition to phosphate, nitrate, and ammonium, it is also possible to find ions such as g/L ppb amounts of phosphate, nitrate, and ammonium. Altering the chemical make-up of the lake has the potential to have significant effects on the ecology that it supports. A major disruption to the landscape speeds up the flow of water, uncovers more soil, and reduces the plant cover, all of which contribute to an increase in the rates of leaching and erosion (Pathak & Mudgal, 2005). In addition to delivering micronutrients such as nitrogen and phosphate, urban wastewater and stormwater inputs also contain essential ions such as chloride and potassium, as well as oil and heavy metals from highway and parking lot runoff.

These contaminants can be found in the runoff. The composition of lakes is impacted by a wide variety of chemicals that are produced by a variety of sources, including emissions from automobiles, power plants and industries that burn fossil fuels, and other sources. Ions H+, SO4 and NO3 that are associated with acid rain have received some of the most extensive research.

Mercury is yet another important air contaminant that affects the ecosystems of aquatic environments. When it accumulates in aquatic food webs, there is a possibility that it will be hazardous to human and animal health. In lakes with hard water, the ions calcium and magnesium are present in very large quantities, whereas, in lakes with soft water, these ions are found in relatively low concentrations. There is a significant correlation between the amount of bicarbonate present and the hardness ions, most notably Ca+2 ,in particular. The ionic concentrations of the lake affect the lake's ability to degrade pollutants and keep nutrients dissolved in the solution.

As an illustration, calcium carbonate in the form of marl has the potential to precipitate phosphate from water, so eliminating this necessary component. The total amount of ions that are present in water is referred to as the water's total dissolved solids (TDS). The (TDS) content of water, as well as the relative amounts or ratios of the various ions in the water, can influence several different chemical processes. Not only is the introduction of the invasive zebra mussel causing significant shifts in Lake Erie's ecosystem, but the calcium needs of the lake also highlight the region's unique significance. Due to the low calcium concentration of Lake Superior, it would appear that this invader has been mostly unsuccessful in its attempt to colonize the lake (Owusu-Boateng et al., 2022). However, the lower St. Louis River as well as the port of Duluth-Superior may be susceptible to invasions of zebra mussels.

During the spring and summer, when high light radiation increases photosynthetic activity, there is a surge in the level of biological activity. In addition, during the summer months, the majority of lakes that are found in temperate zones develop a stratification. Chemical patterns are formed in water as a consequence of events such as the stratification of temperature and the actions of living organisms. Because of photosynthesis and the movement of air through the epilimnion, the carbon dioxide concentration is rather high throughout the summer.

2.5 ZONATION

A vast variety of different biological communities are connected to the physical structure of a lake in a typical setting. Light can penetrate silt, which in turn encourages the growth of aquatic plants in the littoral zone along the coast. At this depth, the light levels are often less than one percent of what they are closer to the surface. In addition, the euphotic zone of the lake, which may be found everywhere from the lake's surface down to its depths, is denoted by a light level of one percent. In the epilimnion of the majority of lakes is where you'll find the euphotic zone. In translucent lakes, photosynthesis can take place in the hypolimnion, which is located below the thermocline. Even though the mixed layer, also known as the epilimnion, in western Lake Superior near Duluth, Minnesota, only extends to a depth of around 10 meters, algae photosynthesis and growth can continue to at least 25 meters deep during the whole summer.

Ultra-oligotrophic even though the summer mixed layer only gets to be around 10 meters deep in Lake Tahoe, the lake has a lengthy history of algal blooms that have reached depths of more than 100 meters. Since 1960, Lake Tahoe has seen a significant degree of clarity loss as a result of an increase in the number of algal blooms as well as sediment imports from streams and coastal erosion, all of which have contributed to the lake's decrease. In addition to serving as a source of food and a substrate for crustaceans and algae, the higher plants that grow in the littoral zone are responsible for the formation of a habitat that is unique from the open ocean for a variety of fish and other animals (Ohtaka et al., 2011). Within the limnetic zone, light is unable to penetrate the ocean floor. The benthic zone, often known as the bottom sediment, is covered with a thick layer of different kinds of organisms. Organic-rich sediments may have a surface layer that's somewhere between 2 and 5 centimeters deep thanks to the activities of benthic organisms.

The benthic zone is home to a vast range of different kinds of invertebrates, such as the larvae of dipteran insects and very small crustaceans. It is affected by factors such as the amount of organic matter present in the sediment, the quantity of physical structure, and the pace at which fish are consumed under specific conditions. Sand, which is deficient in organic materials, offers minimal protection to predators and provides little cover for them. Sand, due to its instability and lack of nutrients, inhibits the development of plants in soils that are classified as sandy. Rocky bottoms are home to a wide range of ecosystems, including organic "ooze" pockets and algal substrates. Benthic creatures can obtain a great deal of food on a flat, mucky bottom; nevertheless, this type of environment offers less protection and a narrower range of structure types.

It is possible, in theory, to utilize food chains and webs to investigate the biological communities that exist in lakes to acquire a more comprehensive knowledge of the functioning of the ecosystem. The ecological pyramid is the most effective visual representation of the species hierarchy that may be found in an area. Herbivores, planktivores, and the few carnivores in the ecosystem all get their food from the same enormous number of primary producers. There are a great number of omnivorous animals, and this is because they do not cleanly fit into any of the three primary levels of the food chain. In addition, the levels of consumers experience significant shifts during their work. A fish larva could first consume food particles such as algae, bacteria, or trash in its early stages of development. It is possible that in the future it may start feeding on bigger zooplankton and even juvenile game fish (Naik et al., 2012).

Energy and nutrient fluxes are useful tools to employ when illustrating food webs. Although algae and plants are typically the first steps in the process of maintaining life, one must also ensure that they receive an adequate amount of nourishment. Even though sugar has a high caloric content, our bodies cannot survive only on sugar despite what our children may tell us. Photosynthesis and respiration are the two most essential processes for life to continue, and they are essential in lakes just as they are on land. Sunlight is the catalyst for the transformation of inorganic, nonliving elements into organic, living plant tissue in a green plant. Photosynthesis is the process that macrophytes and algae in lakes employ to create oxygen. They are the primary producers because they supply the majority of other species with the organic components that are essential to their continued existence and development.

The oxygen content of the lake's air is increased as a direct result of photosynthesis, which also produces carbon dioxide. In places with high levels of photosynthesis, such as during an algal bloom, it is feasible for the water to reach a state known as super saturation. To put it another way, if the water is not given the chance to reach a state of balance with its surroundings, the concentration of oxygen may become dangerously close to being completely saturated. It is essential to take into consideration the fact that the temperature of the water affects the saturation value. When it comes to the retention of oxygen, colder water is superior to warmer water. During the stratification process, photosynthesis is the sole source of oxygen that can be found in the deeper parts of the lake (Mustapha, 2011). If light manages to penetrate the thermocline, only then will this be possible. In lakes where the thermocline cannot be seen, there is no other source of oxygen coming from within the lake.

The survival and expansion of algae and higher plants are impossible without the presence of mineral fertilizers and sunlight. The great majority of blue-green algae are unable to flourish in anoxic water, although there are a few notable exceptions. Both the atmosphere and the watershed contain a significant amount of carbon dioxide (CO2), which results from the breakdown of carbonate rocks, such as limestone, and the inhalation of organic material by all of the lake's inhabitants. While algae obtain their dissolved mineral nutrients from the water, higher plants get theirs from the sediments. Higher plants get their dissolved mineral nutrients from the sediments, but algae get theirs from the water. Phosphorus and nitrogen are commonly the most important nutrients since they are typically present in amounts that are low enough to inhibit the growth of algae unless pollution is present. Phosphorus and nitrogen are found in most water sources.

There are adequate concentrations of a great number of other important minerals, including the principal ions and trace metals. It is not always enough for diatoms and a few other species of algae to just have silicon in the environment. Sulfur is a mineral that is required for the survival of all life on Earth, and lakes are a common supply of this element. The processes of photosynthesis and respiration are essential components of the food chain, which include plants, animals, and even microbes. The majority of food webs are highly complicated, and a lake setting might have hundreds of different species living in it. For there to be enough energy for the few enormous fish, there needs to be a large food base consisting of primary producers. The higher up the food chain you go, the less energy you have (M Ongaratto et al., 2018).

These plants have two potential outcomes: either they will be consumed by primary consumers, which is the second trophic level, or they will die and degrade. Zooplankton, larval fish, and other types of invertebrates are all components of this food chain, which begins with algae and progresses up to higher forms of plant life. Microscopic fishes are examples of secondary consumers because they feed on larger fishes. Tertiary consumers include a wide variety of organisms, including humans, ospreys, and giant fish. As a consequence, photosynthesis results in the production of energy and nutrients that are transferred further down the food chain.

Through the process of recycling, particular nutrients make their way back to the beginning of the chain. The release of the sun's energy occurs through a process known as respiration, which is the oxidation of organic matter. Both plants and animals are dependent on their ability to breathe and take in oxygen to continue living. Microbes consume the vast majority of the oxygen that is available in an environment during the process of decomposing organic waste. Decomposers are organisms that not only bury waste from plants and animals but also recycle nutrients that are used in photosynthesis. The number of extinct animals that can be found in a lake significantly outnumbers the number of living ones. The organic component of dead stuff is referred to as detritus, and it can take the form of minuscule plant and animal fragments or organic material that has been dissolved. In recent years, zooplankton that feeds on debris and the bacterial community that is associated with it has emerged as a significant new food source in lakes. This pathway was previously unknown.

The investigation of lakes' primary production accounts for the vast majority of the work that is being done in the field of limnology today. It is necessary to have an understanding of plant development to comprehend the evolution of lakes and the myriad of organisms that inhabit them. The most significant danger that lakes face comes from the excessive growth of primary producers, which is caused by inadequate management of land use that results in excessive nitrogen inputs. Because of this, additional research into these species is needed (Medeiros et al., 2012). The littoral zone is an ideal environment for the growth of macrophytes, which are defined as aquatic plants that have both roots and the ability to float.

It's possible that there aren't many macrophytes in the lake because the bottom is too rocky or sandy for plants to cling to, the wave action is too intense, or the water is too deep. Even in relatively shallow areas, sunlight might not be able to reach the water's surface if there is a significant amount of algae or silt covering the bottom. Algae are the world's second-largest primary producer, behind only plants. They are available in a diverse assortment of shapes and sizes to choose from. The vast majority of it is made up of relatively small colonies or cell filaments. Periphyton is algae that are attached to plants, as opposed to phytoplankton, which is free-floating in the water. The phytoplankton that lives in open water thrives there because they can get their nutrients from the water and get their energy from the sun. If there are a significant number of them, they will become cloudy, take on a greenish or brownish hue, and lose their clarity.

It is possible to categorize the pigments that phytoplankton use in the process of photosynthesis into a variety of different groups. In addition to chlorophyll, algae contain a diverse array of pigments, which enables them to take in light of varying wavelengths. Chlorophyll is the pigment responsible for algae's green coloration. The green algae are the higher plants' closest living relatives. There is a large amount of variation across all of these characteristics, even between species that belong to the same general genus. Diatoms are part of a wider group of golden-brown algae that also include dinoflagellates and chrysophytes. Silica cell walls are the most distinctive feature of diatoms and chrysophytes. Two sections come together like a petri dish in the cell walls of diatoms (Matta & Uniyal, 2017). Silica scales, spines, and lyrics, shells in the form of vase-like structures, can be found on several chrysophytes. Because diatoms cannot move, water turbulence is essential to their survival.

The flagella of Chrysophytes allow them to move across the water column. Several other significant algal assemblages have motile forms. Algae flagellated with golden-brown flagella are called dinoflagellates. Fast-moving cells are capable of placing themselves in ideal light and nutrition conditions. For example, in the depths of the water column or during seasons when sunlight is limited, cryptomonads, which are flagellated algae with pigments, may photosynthesize even in low-light conditions such as those described above. Cyanobacteria is the accurate term for blue-green "algae" because they are bacteria except for their chlorophyll-based photosynthesis. They are the focus of the majority of research and management efforts in eutrophic lakes because of their propensity to cause nuisance blooms.

Aquatic ecosystems depend on the primary producers represented by blue-green algae, despite the negative connotations associated with these organisms. Toxic or discomforting effects can be generated by blue-green algae due to the range of properties they possess. The buoyancy of some blue-green creatures may be adjusted. It all relies on the amount of light and food that is available. To acquire nitrogen from the surrounding water, all plants, including algae, must first absorb either nitrate or ammonium. Ammonium may be produced inside the cells of some blue-green algae by a process known as nitrogen fixation, which involves the transfer of molecular nitrogen from the surrounding environment and the water. It is possible for them to quickly develop when other forms of nitrogen are reduced to the point where they prevent other algae from being able to exist (Benito et al., 2020). Blue-green algae are well-adapted to phosphate deficiency when there is sufficient phosphorus available due to their capacity to absorb and retain more phosphorus. This allows them to thrive in environments with low phosphate levels.

In contrast to green algae and diatoms, blue-green algae, also known as cyanobacteria, should not be ingested by humans. Grazers could have a hard time digesting the enormous cell colonies that are coated in a gelatinous framework that is generated by some blue-green algae. These compounds can also give the grass a "terrible flavor," which discourages animals from eating it. As a result, they have an edge over other types of algae in terms of the ability to escape being consumed and the nutrients they contain. *Aphanizomenon flos-aquae* is capable of regulating its buoyancy and fixing nitrogen; in addition to these abilities, it may also cause algal blooms and form enormous colonies that are inedible. Blooming plants come in a wide variety of genera, but two of the most frequent are Microcystis and Anabaena. They are known to create poisons that have the potential to kill cattle, and it was just recently discovered that they also produce neurotoxins and hepatotoxins; this leads us to believe that they may pose a far higher threat to public health than we had previously believed they would.

For the majority of monitoring systems, doing a microscopic count of all of the hundreds of species of algae that may be found in the water column of a lake would be prohibitively expensive and logistically impossible. Additionally, it is possible that the majority of specialists won't be able to identify a sizeable portion of the algal biomass present in several lakes. In contrast, measuring the quantity of chlorophyll, which is an accurate indicator of the amount of algal biomass present in water, is a far less difficult task (Cardoso-Silva et al., 2018). Chlorophyll is a green pigment that plants utilize to transform the energy that they get from sunlight into chemical energy. To determine the level of chlorophyll that is present in a given sample of water, it is necessary to first strain out any particles that are larger than one millimeter in size using a very small glass-fiber filter.

After determining how much chlorophyll this material contains, the chlorophyll is next extracted with the use of a solvent. Since long years ago, the chlorophyll content and Secchi depth of lakes have been used to evaluate the algae that live there. The Secchi depth may now be evaluated more quickly and at a lower cost than ever before. It is essential to remember that non-algal particulate particles, such as silt from stream flow or resuspended bottom sediment, may affect Secchi data. This is something that must be kept in mind at all times. The tea-colored tint of certain lakes, which is created by the decomposition of organic waste from peat bogs, may also affect Secchi depth measurements. This effect is not always present. Even if chlorophyll-a concentrations are measured, it is still advised that periodic microscopy of the lake's algal population be performed. This is because the species mix may impact the decisions that are made about lake management.

2.6 CLASSIFICATION

It is a natural human tendency to classify things. It is beneficial for relationship imagery and communication. Size-based comparisons allow us to claim that certain items are massive while others are little. Someone will certainly ask, "How big or small?" Depending on its size, we respond with "quite massive" or "very large." Certain classification schemes are supposed to be constructed on a scale. We split the group's bylines to avoid confusion since these limitations do not indicate what is large and what is little. Complications arise when it is claimed that the lines are not in the proper area. We have reached a sensible consensus on where the borders should be drawn for the benefit of communication, at least within the scientific community. There is, however, much disagreement. Categorization systems with clear boundaries and no continuity are simpler to construct. In-plant or animal taxonomy, a robin is never a blue jay. However, the elements being classified may not necessarily fit under the same umbrella in different categorization systems (Lund & McCullough, 2008). When it comes to lake classification, three techniques come to mind; others may exist, but we'll focus on these three.

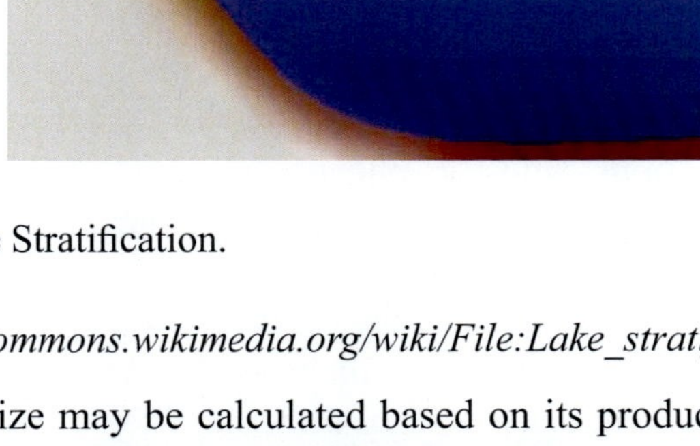

Figure 2.5: Lake Stratification.

Source: https://commons.wikimedia.org/wiki/File:Lake_stratification.png

The lake's size may be calculated based on its production or, as some prefer to call it, its relative nutrient density. Most riparian landowners are aware of this trophic foundation for classification. Oligotrophy and eutrophy are components of a constantly evolving system. Another technique is based on the seasons and the amount of mixing in the water of a lake. This mixing approach is based on the idea that a lake may or may not fall into a certain group. A third method for keeping fishermen happy is to increase the number of fish in lakes. A lake, for example, may be referred to as a cold water fish lake. This indicates that the lake has trout. Although there are significant overlaps in the fish community system, it has enough generalizations to be effective for fish management (Leipe et al., 2014).

2.7 TROPHIC ANALYSIS

The quantity of nutrients or lake production serves as the foundation for trophic classification. It ranges from low nutrient concentration lakes to lakes with high nutrient content but poor quality. As previously stated, this spectrum includes oligotrophic and eutrophic lakes. Drawing borders and applying limitations along this continuum to generate new categories has become common practice. Even the most fundamental categories have exceptions to the norm. But let's start at the other end of the spectrum when nutrients are few. Because the nutrients required for plant development are so low in these bodies of water, oligotrophic lakes produce little. Oligotrophic lakes have low organic matter concentrations, with little phytoplankton, attached algae, macrophytes, bacteria, and fish present.

Corn grows more slowly on sandy soil. Plankton and other species can be found in a variety of forms, however, they are few. Giant fish are not uncommon, albeit there aren't many of them. Because organic molecule production is limited in oligotrophic lakes, organic silt does not settle. To put it another way, when organic food is scarce, we detect fewer microorganisms in the environment. Furthermore, because there are fewer plankton and bacteria in deeper waters, less oxygen is taken.

An oligotrophic lake is oxygen-rich from the surface to the bottom. Clean, algae-free water; phytoplankton with low chlorophyll levels; low nutrient concentrations, mostly phosphorus; and minimal suspended algae are examples of these. These are only a handful of the chemical characteristics discovered thus far. Oligotrophic lakes have sandy or stony bottoms and have similar watersheds, resulting in limited fertilizer inputs into the lakes. Oligotrophic lakes are distinguished by their clear water and absence of flora, which makes fishing difficult. They are usually deep and full of cold water (Larsen & Alp, 2015).

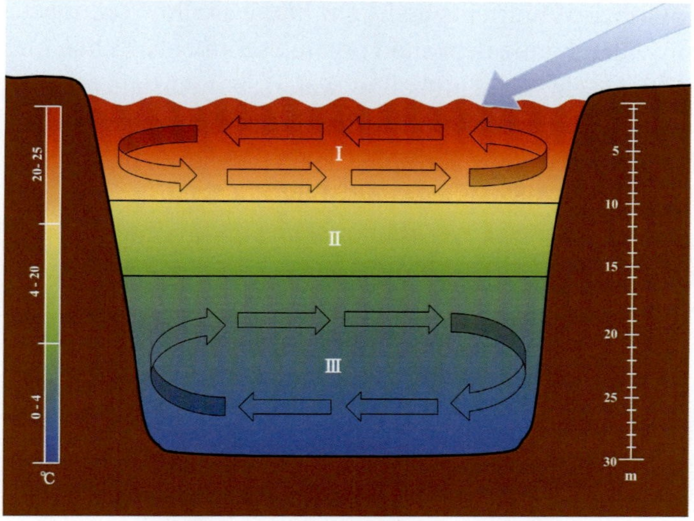

Figure 2.6: Lake Stratification.

Source: By Mbrookings19 - Own work, CC BY-SA 4.0, https://commons.wikimedia.org/w/index.php?curid=69783069

Because of their extensive use and scarcity in highly populated regions, they are no longer categorized as oligotrophic. They are rare in agriculturally productive areas, where fertile soils keep drainage water from reaching oligotrophic lakes. The Top Peninsula and the upper third of the Lower Peninsula are home to Michigan's oligotrophic lakes. In every way, eutrophic lakes are the polar opposite of oligotrophic lakes. They may produce a large amount because of the number of plant nutrients. Secchi disk measurements become increasingly difficult as phytoplankton levels rise. Zooplankton, which minnows and other small fish devour, is also abundant in these lakes.

As a result of the food given by smaller fish, larger fish can flourish. In general, organic matter output is comparable to maize production on rich soil. As the majority of biological waste sinks, organic silt collects on the bottom. This silt, in turn, supplies food for a wide range of microorganisms. In the deepest parts of these lakes, plankton and bacteria can entirely deplete the oxygen. One of the characteristics of eutrophic lakes is midsummer oxygen deprivation. Because of the abundance of phytoplankton, the eutrophic lake typically has chlorophyll values of 14 mg/m3 or higher.

The normal phosphorus concentration is 80 mg/m3. Eutrophic lakes are often shallow and densely vegetated. Weed beds are abundant in these lakes due to the number of nutrients and light in the shallow sections, as well as the "soil" provided by the deposited organic sediments. When plankton and benthic species are abundant in shallow waters, they sustain massive numbers of fast-growing fish. The bottom two-thirds of Michigan's Lower Peninsula are densely populated with eutrophic lakes. As a result, oligotrophic and eutrophic lakes are at the opposite end of the eutrophic continuum. Humans, on the other hand, feel that a lake is more than oligotrophic and not even close to eutrophic (Dar et al., 2020). Furthermore, we show that the oligotrophic and eutrophic groups have a transitional time. Even as lakes age, they absorb nutrients and sediments that contribute to eutrophication.

Physical lake filling and chemical lake water enrichment are both eutrophication procedures that can last for thousands of years. Human activities in lake drainage basins can cause chemical enrichment of lake water over time, leading to cultural eutrophication within two human generations. A mesotrophic lake is located between oligotrophic and eutrophic water levels. The Mesotrophic Lake, for the most part, stands in between an oligotrophic and a eutrophic stage in the ecosystem. As a result of insufficient plankton development, organic matter accumulates and oxygen is lost at the ocean's bottom. At the bottom of the well, oxygen may not have been completely exhausted. Secchi disk depths, as well as phosphorus and chlorophyll contents, indicate that the water in oligotrophic and eutrophic lakes is clean. Most mesotrophic lakes feature a few scattered weed beds with few weeds within. Despite their superb fishing, mesotrophic lakes cannot keep up with demand as well as eutrophic lakes.

CHAPTER 3

HISTORY OF LIMNOLOGY

CONTENTS

3.1 BACKGROUND

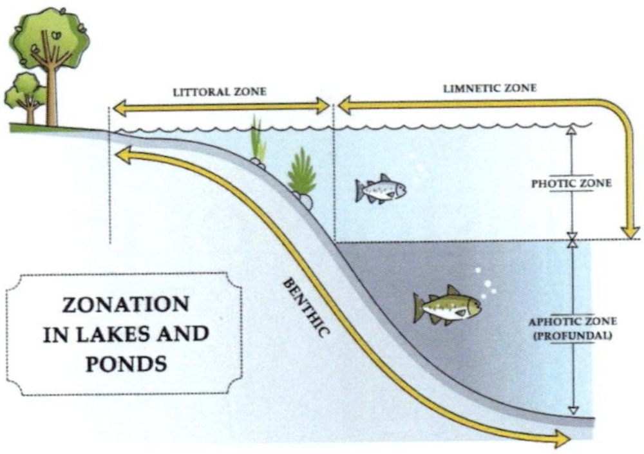

Figure 3.1: Zonation in lakes and ponds.

Source: https://www.shutterstock.com/search/limnology

Limnology is defined as the integrated study of the geological, chemical, physicaland biological properties of waters found within continental boundaries. Limnology differs from oceanography in that oceanography studies open ocean waters. There are different kinds of inland waters found in continents that include rivers, wetlands, reservoirs, lakes, springs, groundwater, and estuaries. There are different areas of specialization for limnologists. Limnology specializations include landscape limnology, ecohydrology, aquatic biogeochemistry, freshwater ecology, etc. One of the primary objectives of limnologists is in providing guidelines for controlling water pollution and improving water management. Limnologists are greatly involved in studying ways to protect wildlife living in rivers and lakes (Kumar, 2022). There is a group of limnologists who are working on the construction of artificial wetlands. Some of these wetlands could serve as habitats for plants and animals or act as water filtration systems to help purify contaminated water.

Limnology can be considered a relatively new academic subject Francois Alphonse Forel, a Swiss physician, is considered the founder of limnology, his earliest observation in the field was dedicated to the study of the relationship between the biology, physics and chemistry attributes of Lac Leman (Lake Geneva). The environment he grew up in greatly affected the development of the field of limnology. *Circa* 1892 he coined the term "limnology", derived from the derived from the ancient Greek word λίμνη (limne) meaning lake or pond and Latin limnaea - thing pertaining to a marsh.

3.2 FATHER OF LIMNOLOGY

François -Alphonse Forel is the father of limnology. He was born on 2nd February 1841. He was a Swiss physician and scientist greatly involved in the study of lakes. It is said that he pioneered the study of lakes and is considered the founder and father of limnology. Limnology is defined as the study of bodies of freshwater and their physical, chemical and biological features. A keen look at his life from early childhood to adulthood shows that the environment he grew up in greatly contributed to the development of limnology. Forel was born in Morges, Switerland around Lake Geneva, a lake that greatly influenced his research. Forel's father, Francois Marie Etienne Forek was a well-respected historian. He was also a jurist and had a very strong influence on Forel's life. This is because, while Forel was still young, he would (Forel's father) would involve him in some archeological studies (Knight et al., 2015). Other than his father, other members of Forel's family members were also respected scientists and scholars. Forel's involvement in some of the studies at a young age greatly influenced his decision and was a form of training to help him gain all the needed knowledge and skills in the field of research.

Figure 3.2: François -Alphonse Forel is considered the "father of limnology".

Source: By Leo Wehrli - This image is from the collection of the ETH-Bibliothek and has been published on Wikimedia Commons as part of a cooperation with Wikimedia CH. Corrections and additional information are welcome., CC BY-SA 4.0, https://commons.wikimedia.org/w/index.php?curid=68552667

His education also allowed him to gain the much-needed knowledge in research. He began his education at the college de Morges. He later left Morges for secondary school in Geneva at the Gymnase de Genéve. It is in his secondary school that he got to study medicine and natural sciences. After completing his secondary education, he attended university at Academies de Genéve. He got his Bachelor of Letters and Bachelor of Physical and Natural Sciences. He left his home country when he turned 18. His move was to allow him to attend the Académie Fe Montpellier in Montpellier, France. He took two years to study at the university and took some time after that to study medicine on his own (Knott et al., 2002). This he did while in Paris, France. Forel then moved to Germany to study at the University of Würzburg. When he was 24 years old, in 1865, he got his doctorate degree in Medicine and Obstetrics.

Figure 3.3: Lake Geneva played a central role in François -Alphonse Forel's studies.

Source: By Schnäggli - Own work, CC BY-SA 3.0, https://commons.wikimedia.org/w/index.php?curid=22892071

This allowed him to enter his professional and work life where soon after getting his doctoral degree, he returned to Lake Geneva in Switzerland. While there, he began to study the lake. His studies encompassed zootomy, biology, chemistry, natural history, economics and physics. This eventually led to Forel becoming a science professor at the University of Lausanne. The university is also known as the Academy of Lausanne and was located near his home at Lake Geneva. While at the university, he taught physiology, anatomy and histology. This, however, was not his passion as his real love was the lake. All studies and investigations done with regard to water circulation, chemistry, biology, sedimentation and their interactions were very useful as they later created the foundation of a new discipline (Kazanci et al., 2004). He later named all his activities as limnology which is in analogy with oceanography. The term limnography is at times confused with alimnology. A limnograph is used in measuring water levels in lakes.

While conducting research, his chief work was the publishing of Le Léman in three volumes between 1892 and 1904. The published volumes covered different topics that included the definition of limnology, the geographic settings, climatology, hydrology, bathymetry and geology of lake systems. The monograph described the lacustrine hydraulics, temperature stratification, chemistry, acoustics, optics, seiches, waves and currents. The monograph was cemented by his publication of a handbook on the topic. During his studies, he discovered the phenomenon of density currents in lakes which allowed him to explain seiches which is the rhythmic oscillations observed in enclosed waters. Forel is also known for his work on the optical phenomenon of the Fata Morgana. The Fata Morgana can be seen when looking out across Lake Geneva and other places.

Forel worked with other researchers such as Wilhelm Ule. They both developed the Forel-Ule scale. This scale is used when evaluating the color of a body of water. He also worked with researchers from different fields such as the Italian seismologist Michele Stefano de Rossi. Their collaboration led to the development of the Rossi-Forel scale used in describing the intensity of an earthquake. He built a legacy before he died. Before his death in 1912, he had gathered a total of 288 academic publications. He had amassed knowledge in lakes. For this reason, he was referred to as the Faraday of lakes. Forel was well regarded by the Scottish scientific community. He was elected as an honorary fellow of the Royal Society of Edinburgh. Because of his great achievements, the institute F.-A. Forel of the University of Geneva was named after Forel. A mountain at Spitsbergen, Foreltinden was also named after him. The same applies to the Forel station in Maule, Chile (Jeong et al., 2008). The station was named after him as he lived close to the station. A submarine used in making scientific measurements in Lake Geneva was also named after him.

3.3 DEVELOPMENT OF LIMNOLOGY

While conducting studies on lake systems, Forel published a three-volume publication. It is noted that in the preface to his three-volume treatise on Geneva, he was struggling with the question of what name he should use for his integrative approach to lake science. As his studies encompassed various aspects of the lake such as the lake's morphometry, underwater light, mixing of patterns and sediments, the challenge of the most appropriate name may have developed. He also discussed studies related to animal, plant and microbial ecology in the ocean. During his studies, he noted that various environmental characteristics of Lake Geneva were closely interrelated (Islam & Chowdhury, 2013). This led him to view the lake as a system that was intimately tied to its surrounding catchment. Through his interactions with his father who was a historian and local jurist, he was able to identify the lake as a precious resource useful in ensuring the good health and wellbeing of large populations of people who lived around Lake Geneva. He saw that the lake would affect people in the past and present.

During his studies, he also quantified the lake fishing economy in a manner that foreshadowed the modern concept of ecosystem services. A general overview of Forel's research interest in Lake Geneva was too large to classify within a specialized subject area. This also made it difficult to try and insert his interests into any existing integrative discipline. All these challenges were explained in his preface. In his preface, he stated there he wanted to achieve a generalization or an overview of all detailed facts such that each specialized data from a given study could be supported by data from other studies. The theme of his description in his publication was partly terrestrial that could cause the subject to be considered Geography. The challenge however was that the geography of water is referred to as oceanography. The discipline would otherwise be referred to as fresh water oceanography. Such a name was not suitable for the study as no matter the size of the lake, it is not an ocean. He further stated that the limited expanse gives each lake a proper character that is very different from the unlimited expanse of the vast ocean. For this reason, Forel decided to come up with a new word for a new discipline that fits his interest. The term "limnology" was thus coined.

Figure 3.4: In Forel's studies, much focus was placed on the melting of glaciers on Lake Geneva.

Source: https://commons.wikimedia.org/w/index.php?curid=5541451: By The copyright holder of this image, Christopher Down, allows anyone to use it for any purpose, provided that the copyright holder is properly attributed in a position close to the image. Redistribution, derivative work, commercial use, and all other use is permitted. If this license is too restrictive please e-mail me. Use outside of these terms requires permission.

All the accomplishments of Forel are usually recognized globally in that in the early 29th century, he has been internationally recognized as the founding authority of lake science. J.G. Needham and J.T. Lloyd wrote the first English Language textbook on limnology. In their book, they noted the profound influence of Forel also referred to as the father of Limnology.

Some of the achievements highlighted in their book was the Forel was the first to study lake intensity after Morden methods, he made Swiss lakes the best known of any lakes across the world. They also recognize the greatest work of Forel which is the Le Léman which is a monograph on Lake Geneva (Ishida et al., 2020). This monograph is considered a masterpiece of limnological literature. Forel is also the first to develop a comprehensive plan for the study of the life of lakes and the lake's environmental conditions.

All the tributes accredited to Forel in Lloyd and Needham's book does not highlight all his achievement in the development of limnology. Some of the lesser-known contributions of Forel in the development of Limnology include the seminal insights into lakes such as hydrologic optics and lake water color, microcosms, the structure and function of aquatic food webs including microbial decomposition processes, the nature of pelagic and littoral communities and the organic carbon cycles. He also made various observations with regards to the ecology of invasive species, he made noted the intimate connection between freshwater resources and human society and the role of human beings as a powerful biotic component of lake systems.

Some perspectives of Forel with regards to Lake Geneva are currently used as a reference point in Limnology today. Some of the perspectives include the surrounding of the lakes when coupled as a physical-chemical-biotic-human system (Domisch et al., 2015). This perspective is said to be relevant today in light of current environmental challenges and the associated rapid shifts in ecosystem services on a global scale.

3.4 FOREL'S EARLY CAREER DISCOVERIES AND NETWORKING IN THE DEVELOPMENT OF LIMNOLOGY

At the age of 29, Forel was a professor and had fairly light teaching load at the Academy. He had about 3 hours of lectures per week. This means that Forel was left with plenty of time to conduct research. He also benefited from the fact that he had the academic freedom to choose his subject. Forel acknowledged the fact that setting up a science program in physiology, histology and anatomy was there the most logical set of options given his teaching obligations in the subjects he taught. While in the university, Forel decided to make Lake Geneva his laboratory. He then proceeded to make his subjects of research to be all aspects of the ecology and physics of the lake (Hodgson et al., 2001). There were those who had issues with the fact that he was dealing with various aspects rather than focusing on one. This included his former professor Albert Kölliker. He voiced his concerns during his visit to Lausanne in 1871. They suggested that Forel focus his studies on a zoological topic. However, this did not dissuade Forel.

This is because, during his first years of research, he had already gathered very exciting results which reinforced his plans for his career in lake science. He had developed an interest from his earliest days of working with his father. He had a passion for working with lakes and learning more about it. He was so passionate about it that he was later on invited by an editor to write a textbook on Limnology, an opportunity that he had earlier declined stating that he could not remain calm and impartial about his subject. This is all included in the preface of the textbook. He stated that his relation to limnology as a subject was way too personal and subjective for him to be able to give an objective presentation of facts. Like Forel, the editor was not persuaded by Forel's objection to writing the book allowing the project to be actualized. One achievement of Forel was the accidental discovery of deep-living nematodes in the lake.

Figure 3.5: Deep-living nematodes were found in Lake Geneva.

Source: By Trisha J. Brock, John Browse, Jennifer L. Watts - Brock TJ, Browse J, Watts JL (2006) Genetic Regulation of Unsaturated Fatty Acid Composition in C. elegans.PLoS Genetics Vol. 2, No. 7, e108 https://dx.doi.org/doi:10.1371/journal.pgen.0020108, CC BY 2.5, https://commons.wikimedia.org/w/index.php?curid=1676254

Forel made a visit to Germany and returned after two years in April 1869 which led to him making an observation on the rippling and deposition of sediments on floors of lakes. He used a metal sampling plate to take up some sediment from the depth of 40m offshore from Morges. He then placed a sub-sample under the microscope to examine the nature of the material. It is from this examination that he got to see a nematode swimming about. The nematode species was identified as Mermis Aquatilis. This got him really excited. He had made the discovery that the lake floor of Lake Geneva was not sterile as by other scientists. It also changed his views that lake floors could be the habitat for a specialized community of deep-living animals. He then proceeded to construct a dredge and returned to the lake to gather more samples at various depths (Helm, 2017).

This provided solid data to prove that there is a diversity of species occurring in the sediments. He found species even in the depths of 300m which is the deepest abyssal region of the lake. This led him to start a project that later on achieved enormous success. The success of the project is attributed to the fact that Forel decided to collaborate with other scientists rather than choosing to use the generated data all for himself. The project was later on (in 1874-1879) published as a series of 59 published reports having 19 collaborators. It was later culminated in his 1885 publication called La faune profonde Des lacs suisses also known as the Deep Fauna of Swiss Lakes. The volume publication is broader than the suggested title. It is comprised of sections on chemicals and physical limnology and descriptions of several large lakes of the world. During his first years in Lake Geneva, Forel made a discovery that contributed to the growth of his career. At that time, the boat harbor Morges had two exits to the lake. There was a much smaller one that was 2m across that allowed exchange of water while the other allowed the passage of boats (Hassan et al., 2018). This was an area for study in that from 7 May 1869 to 11 May 1870, he took a series of careful measurements on water flow through the small aperture. He discovered that water currents oscillate in the direction whereby it would flow into the harbor and layer on the flow would reverse into the lake.

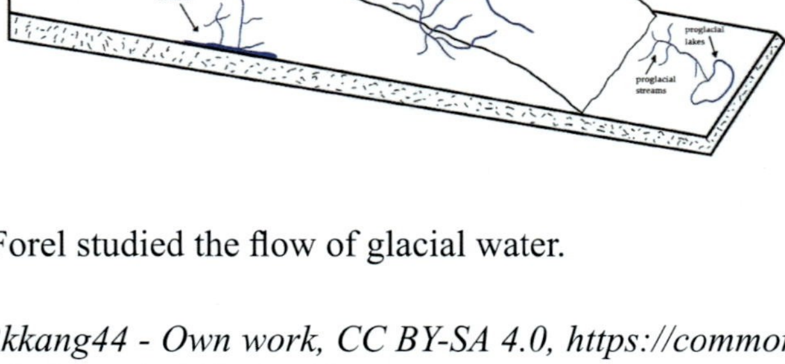

Figure 3.6: Forel studied the flow of glacial water.

Source: By Rkkang44 - Own work, CC BY-SA 4.0, https://commons.wikimedia.org/w/index.php?curid=103764718

From the study, he got two insightful revelations from the pattern of oscillation flow. In his first revelation, he surmised that the pattern of oscillation of flow was somehow connected to the periodic rise and fall of lake level noted around the lake more so in the city of Geneva for a number of years. This in the local Swiss French dialects was referred to as the seiche which means rocking back and forth. In his second revelation, he concluded that the phenomenon was as a result of standing waves that was propagated throughout the lake. He proceeded to publish a paper on this topic and continued to amass data from observations he made and from other sources around the lake.

Forel's interest in physical limnology was stimulated and encouraged by his former professor of physics and mathematics who had an interest in geology, glaciology and meteorology. Forel constructed the first series of limnographs in 1876. The limnograph is an automated instrument used in obtaining continuous records of lake levels. After the construction of the limnograph, he proceeded to design tide gauges mostly used by oceanographers. These tide gauges were very different from tide gauges used in marines at that time as they had the ability to resolve mm-level fluctuations in water levels. These tide gauges were also able to pick up any vibrations of steam ships passing many km away. In the following years, he worked with other colleagues at Lake Constance and Lake Geneva (Gothwal, 2021). They also worked with colleagues from other areas. During their investigations, they worked towards defining the properties of surface seiches. To get highly accurate data, Forel corresponded his research on this topic with other scientists around the world including those from Japan, Australia, England, France and USA.

He faced a number of challenges including his lack of formal training in differential calculus. He was able to deal with this problem through the network of colleagues he had created. He was able to gain the needed expertise which allowed him to develop a quantitative theory of Seiches. He had one contact at the University of Leipzig who gave him a treatise that was published in 1828 by a family member who was a mathematician by the name J-R Merian. The mathematician applied Lagrangian analysis to a problem pertaining to water movement in a rectangular tank. Forel took the equation and modified it so that it could be used in lake basins. This enabled him to make estimates of the average depth of the lake based on the period of the seiches.

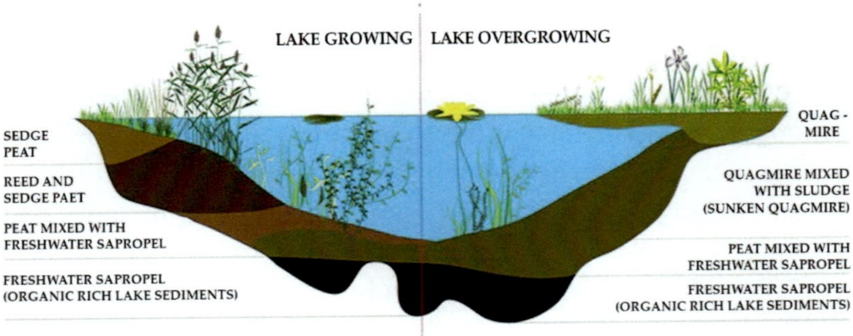

Figure 3.7: Various topics were covered in Forel's publications.

Source: https://www.shutterstock.com/search/limnology

Despite the achievement he made using the equation he was still unsatisfied with the fact that the equation was very complex. This led him to hold further meetings and discussions with other researchers. The issue of the complexity of the problem was dealt with by a distinguished physicist at the University of Glasgow who was able to assist Forel deal with the problem. He wrote a letter to Forel. In the letter, he took the Merian-Forel formula and simplified it for the condition that the size of the lake is very small relative to the length of the lake. He also deleted the minor terms so as to produce an elegant approximation. The budget for the approximation totaled 15000 Swiss francs. This was a factor that contributed to the plan not being implemented. It however still stands as the first comprehensive environmental research on a lake. Its formulation and signing were done 140 years ago. This research is considered the birth of limnology (García-Girón et al., 2020).

At least, the set research agenda for Forel for a number of years resulted in the production of the three-volume treatise on the limnology of Lake Geneva.

3.5 CORRESPONDENCE AND MONOGRAPHS

Forel undertook a 25-year tenure at the Academy of Lausanne where he taught and also did continuous outputs of scientific publications, frequent presentations at scientific meetings and prolific correspondence with colleagues in Switzerland and abroad. It can be worth noting that he was able to make most of his achievements through various collaborations. A good example of his collaboration is one that he did with the renowned

French limnologists Andre Delebeque who is also the author of the classic monograph on French lakes. Their collaboration resulted in the production of the first bathymetric map of Lake Geneva. There are dozens of letters that have been achieved at the Center for Documentation one being the letter he exchanged with John Le Conte who was a professor of physics and former president of the University of California at Berkeley. He was the first limnologist to work on Lake Tahoe.

Figure 3.8: Lake Tahoe was among the Lakes studied during the development of limnology.

Source: By Lara Farhadi - https://www.flickr.com/photos/peruvianprincess/8680540700/in/set-72157633317567718, CC BY 2.0, https://commons.wikimedia.org/w/index.php?curid=38264345

During the development of limnology, one of Forel's greatest fascinations was the interplay between lakes and sunlight at the surface of the lake and in the propagation of light through the water column. Forel was able to establish a standard protocol for the utilization of the Secchi disk. The Secchi disk was a white circular plate invented by Priest Angelo Secchi. The disk was used in measuring the transparency of the Mediterranean Sea. The device was tested extensively by Forel in Lake Geneva a few years after its publication. Forel, later on, applied his standard Secchi disk method when describing the seasonal dynamics of lake water transparency. It was very useful in indicating lake water variations from around 15 m in winter and around 5 m in summer. Forel was able to develop a number of novel approaches in hydrologic optics (Gagan et al., 2018). This included an in situ method that was based on photographic plates used in measuring photochemically active radiation. He also developed a color scale on the basis of mixtures in colored reagents so as to quantify lake color.

Forel corresponded with Le Conte in a letter dated April 9th, 1884. In the letter, he asks why Tahoe was apparently more transparent compared to Lake Geneva. The clearance was attributed to the absence of active glaciers in the Tahoe basin. In his letter, Le Conte also expressed his interest in Forel's high-tech Limnographs in Lake Tahoe which will be used in measuring seiches. Forel developed a detailed discussion and application of an equation where he noted that the equation was appropriate to basins of simple morphometry. He further noted that for complex basins, other formulations were required. Examples of such formulations include those developed by French engineer Paul du Boys who was a colleague of Forel who had recruited into the seiche discussions.

Forel took part in two meetings at Nyon from 1871 to 1872. The meeting was defined by a broad spectrum of research on Lake Geneva. The meetings were formally commissioned by two scientific societies which brought together distinguished scholars from Lausanne and Geneva. The work of all the research brought together culminated in a report that was prepared and signed by FAF as secretary of the commission (Eaton & Kardos, 2015). In the report, there were 14 outlined research topics that were to be addressed which included lake transparency, sedimentology, bathymetry, geomorphology, water chemistry, currents, fauna, archeological features and aquatic flora of the lake.

The document was published in the same year in number 401 of the Bulletin of the Vaud Society of Natural Sciences. It contained a description of limnological methods that would be applied and proposed old civilization of it is difficult for an individual to appreciate the conditions of a new and undeveloped civilization. Much of Forel's work resulted in dozens of reports, scientific articles and notes. His main goal however was to generate a big synthesis. Some of his articles were based on benthic fauna in various lakes including Lake Geneva. All the collected information was gathered in a volume in 1885 and was later on revised and published as a little guidebook to Lake Geneva that highlighted several limnological features of the lakes. An ambitious overview that he had embarked on was completed by all the data he had gathered in his encyclopedic collection of facts about the lake. It was on 2 February 1891 that Forel got to release a complete manuscript of the first volume of the Le Leman- Monographie Limnologique. The manuscript was subjected to 18 months of proofs and corrections by Forel and his colleagues that led to the publishing of Volume 1 in August 1892.

Forel's work is recognized by the definition of limnology as a new branch of science and articulates the vision that he had as a young scientist. He was able to describe Lake Geneva from diverse yet connected points of view. He wanted to work on different problems at Lake Geneva. He wrote his books to help fellow scientists and individuals interested in explaining natural features that they admire and find intriguing.

Figure 3.9: Forel highlighted the effects of human activities on lakes.

Source: https://commons.wikimedia.org/wiki/File:Fishing_down_the_food_web.jpg

While defining the moment for limnology, Forel captured the broad spectrum of end-users of lake science which included the fundamentals of applied limnology. By doing this, he demonstrated that human beings are an integral part of the lake Ecosystem. This theme comes back in great detail in the third volume. There is a great variation in all the volumes he wrote with regards to data. Volume 1 presents a rich compendium of information that includes a summary of field apparatus that can be used in limnological studies, it also includes detailed descriptions of the geographic setting, geology, climatology, hydrology and hydrographic features. In the volume, there is meticulous attention to earlier work such as the history of the development of knowledge on each subject and highlighting some unpublished data sets (Degefu et al., 2014). The volume is completed by a bathymetric map of Lake Geneva which has 25 m isobaths on the basis of 11,955 surroundings. Forel further stated that 4,338 of the surroundings were obtained from a distinguished French scientist by the name Andre Delebecque. He is considered as Forel's limnological counterpart and was based in Thonon which was the opposite of Lake Geneva.

Forel submitted his second volume two years after submitting the firm volume. Compared to all the released volumes, the second volume turned out to be the most important volume. The second volume contained sections of lake hydraulics that included seiches, hydrologic optics and thermal characteristics of the lake. These sections were followed by a short section on lake acoustics which places much focus on the speed of sound in air and water. There is also a chapter on water chemistry. In the latter chapters, there is a chapter on dissolved material in the water of the lake where he clearly links lake chemistry to catchment properties. Some of these properties include the sulfate concentration of 64 mg. This places much emphasis on chalk sulfate in particular (Dyck, 2007). The presence of gypsum in many valleys of watershed more in the large valley of Rhone accounts for the presence of sulfate. He also noted that in some lakes, there were low rates of dissolved organic carbon concentrations which were linked to variations in runoff and rainfall. The main question was about the nature of organic material in lake water. This question was not well answered as there insufficient studies on the topic.

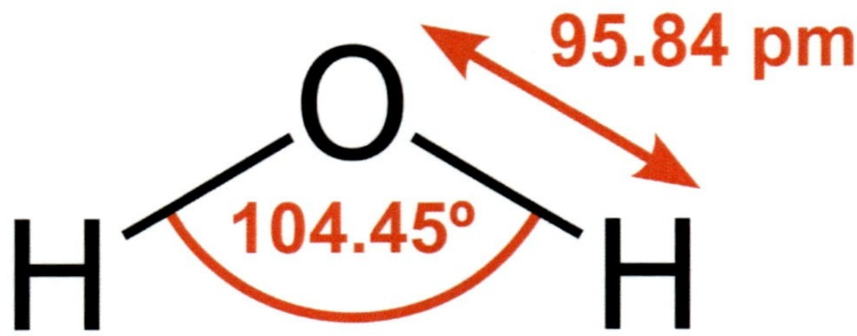

Figure 3.9: Water chemistry was covered in Forel's publications.

Source: By Dan Craggs - This vector image includes elements that have been taken or adapted from this file:, Public Domain, https://commons.wikimedia.org/w/index.php?curid=7916072

The lake was stratified on the sampling date of 30 November 1880. The bottom level was found to be well oxygenated. From this Forel was able to establish an important baseline value for gauging the magnitude of change in Lake Geneva during the period of rapid eutrophication during the 20th century. He then proceeded to present huge amounts of thermal data for the Lake Geneva water column on the basis of dozen profiles from the work he did- in that much of the work was done using high resolution reversing thermometer. He also made use of data generated by his colleagues. From his data, he proceeded to make heat budget calculations for the lake where he was able to establish the importance of density current and convention. He generated a thermal profile that allowed him to define the annual pattern of stratification and mixing. He was also able to identify interannual fluctuations in hypolimnetic temperatures. From this, Forel was able to develop a logical method of classifying lakes on the basis of their stratification regimes (Wu et al., 2021).

Under his classification, he classified polar lakes as those in which ice and cold water overly warmer water during winter stratification followed by mixing in summer. He also defined tropical lakes as those in which warm water overlies cold during the stratified period which is followed by mixing in winter. There are temperate lakes. These lakes are those that have an alternation between tropical and polar modes of stratification separated by periods of mixing. This development was later on acknowledged and adopted by Hutchinson. He did not use the ideology as it was. He replaced confusing terminology in which Lake Geneva was termed as a tropical lake despite its temperate location by terms based on mixing each year. Some of the terms include dimictic, warm monomictic and cold monomictic among others.

In Forel's approach, the goal was to formulate questions and test various ideas. Most of the generated ideas were tested by way of competing hypotheses and encyclopedic synthesis of observations from various sources inclusive of his own work. This was very useful as it allowed rapid insights and rejection of alternatives. A good example is his work on surface seiches where he readily eliminated hypotheses for casual mechanisms such as periodic injection and electrical storm effects of melted waters from glaciers. A significant number of his conclusions have not survived the test of time. This fact was openly welcomed by Forel who stated that science was continuously evolving through the generation of new data and the development of new equipment used in studies. A good illustration is with the case of glacial erosion which is thought to play a great role in carving out Lake Geneva than what was summarized by Forel. There was a bit of reluctance by Forel in accepting the presence of internal waves in Lake Geneva despite evidence present in his own data of thermocline tilting

(Tundisi et al., 2010). He attributed thermocline tilting to the mixing of more dense water from the Rhone at the western end of the lake. Evidence and theory of internal waves from Scotland were discovered later in his career. This was at an unfortunate time when he had a medical problem with his hands which affected his fieldwork on the lake.

As Forel grouped information according to what he was studying, this led to the development of the branches of biology that are as follows:

3.6 HUMAN AND BIOLOGICAL LIMNOLOGY

Figure 3.10: Biological limnology studies look into the biological composition of lakes.

Source: By Original uploader was Yogi de at en.wikipedia - Transferred from en.wikipedia, Public Domain, https://commons.wikimedia.org/w/index.php?curid=3612077

Forel was able to dedicate most of his time to limnology studies after he had retired from the Académie de Lausanne. During this time he was able to collect large amounts of data and information. He synthesized this information as was able to generate several publications. This contributed to him producing his limnological textbook in 1901 and his third volume in May 1904. The 3rd volume was composed of the biological limnology of lakes, the history of human occupation dating back to the Neolithic times, fishery and navigation history of the lake to present data. It can be noted that from his childhood onwards, Forel was fully aware of the significance of Lake Geneva to people who lived around it. This included the many generations of families who lived at Morges. While developing the science of limnology, Forel clearly indicated that human activities in the past and in the present were part of the structure and functioning of lacustrine systems. Details of aquatic and semi-aquatic biota are listed in the 3rd volume of the monograph (Togaki & Katano, 2020). Some of the species found living in the lake were defined as vertebrates, primates and mammals. In the case of human beings, they are not necessarily aquatic species but are considered part of aquatic life by way of activities such as fishing and sailors. There are individuals who live a semi-lacustrine life thereby making mankind almost an erratic species of Lake Fauna.

From his studies, he was able to come to a conclusion that human activities have negative effects on lake water through the building of quays and ports, the channeling of inflows, outflows from dams, discharges into the lake that are products of factories, release of sewers from villages and ashes from steam boats. All of these activities modify the lake environment and therefore has a direct or indirect impact on the biological functions of lakes. It is also notable that human beings exert a more powerful effect on lakes compared to animals in the surrounding of those living in lakes. During his discovery, Forel documented the arrival of invasive animal and plant species in a lake as a result of human interference. A good example is the manner in which the North American macrophyte *Elodea canadensis* was imported and planted in Geneva in 1869. This was done to favor the production of fish. It later on escaped into the lake. Three years down the line it invaded the boat harbor near Forel's residence at Morges. During the first few years, it exhibited spectacular growth and led to the displacement of other aquatic plant species. However, its population was found to stabilize and contract by 1900. This led to the disappearance of all worries about the unstoppable invasion by the American weed.

The American weed went through a phase of exuberant and frightening expansion over a given duration of time after which it settled down as a calmer vegetation type. It was overpowered by other plants growing near it. In his 3rd volume, Forel gives a detailed overview of animal and plant communities of Lake Geneva, all of which are separated according to the abyssal region, the littoral and pelagic zones and the semi-aquatic edge of the lake with its deep-living benthic communities.

After 1869, most of his discoveries on the benthic communities were considered of special interest in that he later on devoted a number of pages towards describing the communities such as the deep living moss *Thamnium lemani*, oscillatorian cyanobacteria, microbial mats and 79 species of animal ranging from protists to fish. He further described the monotonously cold dark climate experienced in the deep region of the lake. He wanted to know whether animals living in these zones demonstrate any general characteristics. He was able to identify several common features and their origins and the evolution of Lake Geneva biota. Forel demonstrated great interest in the deep-living populations of crustacean zooplankton and their nocturnal migration habits (Teffera et al., 2019). He discovered that these animals were able to come to water surfaces during the night and were capable of extensive active movement. He suggested that they come to the surface of the lake during the night because the lake is usually calm at that time. He also noted that they come to the surface of lakes on those nights when the moon does not shine too brightly. During the day, these organisms descend to depths with high levels of obscurity or have less illumination another scientist (Auguste Weismann) also came to the same conclusion in Lake Constance.

3.7 MICROBIAL LIMNOLOGY

In his 3rd volume, Forel gives an introduction to aquatic microbial ecology where he makes some insightful observations that are relevant to the present day. Some of the methods used in microbiology available during his time were extremely limited when compared to those present in the current day and age. Some of the tools used include a microscope which was useful in providing the first indications of Lake Geneva's microbial world. The microscope allowed him to gather all available data on bacterial count from water hygiene records. He noted that despite the many orders of magnitude that was below the modern methods, there were great abundances everywhere. There was no serious analysis to prove that any lake could be completely free of microbes. He also noted that high bacterial population sizes were more so in high surficial sediments, parts of the lake near sewage outfalls, and in waste water all of which had the lowest concentrations in the offshore pelagic zone (Soranno et al., 2010). From his studies, he was able to separate bacteria into functional groups while indicating the essential roles played by such bacteria in the carbon cycle of lake ecosystems through the degradation of organic materials.

Figure 3.11: Microbial limnology studies microbial ecology of lakes.

Source: By Yakuzakorat - Own work, CC BY 4.0, https://commons.wikimedia.org/w/index.php?curid=64811419

The 2nd volume is considered very important compared to others. In this volume, he incorporated hos observations. He found out that clean natural water contains microbes. This however should not alarm common civilians. He also noted that not all microbes will be classified as unhealthy. Contrary to most opinions, most microbes are completely innocent. He covers a chapter titled "The Circulation of Organic Materials". In this chapter, he tries to explain how mineral materials that find their way into the lake are converted into organic materials that are later on distributed among three forms. The first form is that of dissolved organic matter which forms the general reserve stored in waters of the lake. The second form is organic material in the bodies of detritus and dead organisms and the final form is that of organic matter in the state of living animals and plants (Sigala et al., 2017). He further stated that all these compartments are connected and ultimately form substrates for microbial decomposition and demineralization. There are different variations in the successive incarnations in the organic material subjected to lakes as it passes from plants to the animals that feed on it. For example with alga, it is a diatom that can be eaten by a rotifer, copepod, whitefish, otter, pike, cladoceran and human beings.

In natural ecosystems, it is natural for small and weak organisms to become prey for the strong and large which are in turn eaten by the larger and stronger animals. Even in the event that they escape predators, they will not avoid the microbes responsible for decomposition. All organisms are subject to microbes of decomposition either directly or indirectly. This leads to the incarnation in the animal or plant that allows the organic material to pass from an organized state into a state of solution. This led to ecologists such as Raymond Lindeman praising Forel for the brilliant work in providing a brilliant exposition of the general nature of food cycles useful in present day as an inductor account to trophic relationships.

However, Forel was not fully aware of the enormity of the total bacterial biomass and respiration in Lake Geneva. He made rough calculations of the allochthonous input of organic matter which led to the conclusion that a large part of the organic material that enters the lake from its inflows is emitted into the atmosphere. Emission is as a result of gases produced as products of decomposition and putrefaction over organic material either dead or living. The escape process is non-stop into the atmosphere. This can be considered as the first recognition of lakes as heterotrophic ecosystems.

3.8 GEOLOGICAL LIMNOLOGY

This branch of limnology deals with the formation of rivers and lakes. There are many lakes more so in North America that were formed by the great retreat of the glaciers at the end of the ice age. An increase in temperatures led to the melting of glaciers. This led to the formation of holes in soft parts of the solid rocks. Most of these holes were filled with water resulting in the formation of lakes. There are lakes that were formed when tectonic plates pulled away from each other forming rifts called grabens. Very deep lakes are formed when these grabens are filled with water. For example, the deepest lake in the world, Lake Baikal in Siberia, was formed in a graben (Simões et al., 2011). In the case of rivers, they develop a spring in areas of high altitude such as hills and mountains. These springs flow downwards and gather forming rivers. Some rivers are formed as a result of melting snow. Rivers and streams usually flow towards sea level.

Figure 3.12: Geological limnology studies the formation of lakes and rivers.

Source: https://www.shutterstock.com/search/geology

In the study of limnology, those in the field of geology are mostly interested in the topography, size and shape of the watersheds of lakes and rivers. The term watershed is used to refer to land and water areas draining into the river or lake.

3.9 PHYSICAL LIMNOLOGY

Physical limnology as a branch of limnology deal with the study of physical properties of water in rivers and lakes. Therefore studies pertain to changes in water temperatures, light levels and water currents. Water is able to absorb energy from sunlight making water surfaces warmer. The heat from the sun has different intensities which change throughout the year. The amount of heat absorbed during the summer is very high compared to the amount absorbed during winter. For this reason, lakes are highly stratified during summer time. It is characterized by warm water being at the top while cold water is at the bottom. Physical limnology seeks to understand cycles of stratification and mixing as it can be used to explain the biology of animals, plants and microorganisms living in lakes. It also sheds light on the movement of chemicals throughout the lake.

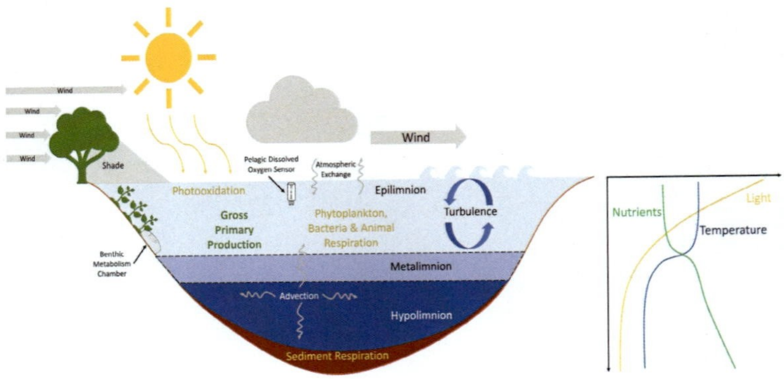

Figure 3.13: Physical properties of lakes are studied in physical limnology.

Source: By Jayzlimno - Own work, CC BY-SA 4.0, https://commons.wikimedia.org/w/index.php?curid=91461560

3.10 LAKE ECOSYSTEM SERVICES

Forel had a theme in mind during the development of limnology. This theme led to the historical importance of lakes to people. His fascination with lakes may be due to the fact that his father was very interested in the palafittes and other buildings found in the littoral zone of Lake Geneva (Shah et al., 2019).

3.11 OTHER SCIENTISTS WHO FURTHERED THE DEVELOPMENT OF LIMNOLOGY

The development of limnology was further championed by other scientists such as the American naturalist Stephen Alfred Forbes well known for his great work in the study of lake ecology. His studies entailed the study of the relationship between organisms and their environment. He published a paper titled "Lake as a Microcosm". This paper is to date cited as an important study of lake ecosystems. The term ecosystem is used to refer to relationships between living and non-living parts of an environment. There is George Evelyn who was a physicist and American biologist. He is known for his great work that helped further the development of limnology at the beginning of the 1950s. George summarized much of the field of limnology in a three-volume text. There is Hutchinson who was extremely influential in bringing modern ecological theories to limnology.

In the current day and age, limnologists have shifted much of their focus towards the integration of ideas from biology, chemistry, physics and geology into understanding rivers and lakes. They have also placed much attention on understanding how human beings affect rivers and lakes as ecosystems.

CHAPTER 4

LIMNOLOGY

CONTENTS

4.1 GEOLOGICAL LIMNOLOGY

Geological limnology, an academic area, studies the genesis of lakes and rivers. Lakes emerged throughout North America as glaciers (huge, slow-moving masses of ice) melted during the closing days of the Ice Age. The softer sections of solid granite were carved off by glaciers as they melted. These depressions became lakes when they were filled with water. Another sort of lake is formed when tectonic plates (moving sections of Earth's crust) split apart and form grabens (rifts). Lakes with unusual depths might form when the fissures are filled with water. Siberian grabens are the source of Lake Baikal, the world's deepest lake, which was generated in the region. In mountainous places with a high altitude, river springs are widespread (Pérez-Fuentetaja et al., 2006). Snowmelt and other streams are known as tributaries flood into rivers as they journey toward the ocean. Lake and river watershed size and form are of relevance to geological limnologists. A watershed is the area of land and water that drains into a lake or river.

4.2 PHYSICAL LIMNOLOGY

Physical limnology deals with the physical qualities of the water in lakes and rivers. This includes oscillations in light levels, water temperatures, and water currents. Water absorbs energy from sunshine, which heats the surface waters. Because the intensity of the sun fluctuates throughout the year, the quantity of heat absorbed in the summer is substantially larger than that absorbed in the winter. During the summer, lakes become stratified or layered, with the warmer, lighter water floating on top of the cooler, deeper water. In the winters, the surface of the lake loses its heat and mixes with the colder waters below. Understanding the cycle of mixing and stratification is vitally critical to understanding the biology of the plants, animals, and microbes that live in lakes and the transport of chemicals across lakes.

4.3 CHEMICAL LIMNOLOGY

Chemical limnology focuses on the cycling of diverse chemical compounds in lakes and rivers. Several components impact the chemistry of lakes and rivers including the chemical composition of the soil in the watershed, the atmosphere (mass of air around Earth), and the composition of the riverbed or lake bottom. In the current day, human activities have had a very substantial influence on the chemistry of lakes and rivers, and chemical limnologists play a crucial role in understanding these effects. For instance, constructing structures near lakes and rivers impacts the erosion (soil wearing away) patterns and determines the kind of pollutants that enter the water. In certain places, precipitation pouring into lakes and rivers creates huge amounts of fertilizers, oils, and heavy metals. The concentration of the hydrogen ion (H+) in water is one of the most critical molecules to evaluate. An ion has a positive or negative charge and the hydrogen ion reveals the acidity (charge) of the water, which greatly restricts which sorts of animals may live in the water (Papa & Mamaril, 2011). Other major components include the sulfate and nitrate ions, which are concentrated in freshwaters as a result of acid rain. Also, the heavy metal mercury (Hg) is a dangerous contaminant that can circulate in the water and impair the health of animals, along with the humans that consume those animals and utilize the lake or river.

Lake Baikal Lake Baikal, which is located in Siberia in the center of Asia, is both the oldest (25 million years) and deepest (1 mile or 1,600 meters) lake in the world. It houses one-fifth of all the unfrozen fresh water in the globe. Because of its antiquity and its isolation from other water bodies, at least 1,200 different animal species and 1,000 varied plant species have formed in it. It is thought that at least 80 percent of these species are exclusively present in Lake Baikal. One of the most interesting creatures in the lake is a seal dubbed the nerpa. It is the lone animal that inhabits the lake. Scientists assume that they migrated to the lake roughly 22 million years ago from the Arctic Ocean. It is expected that the nerpa population in the lake is approximately 100,000. They eat fish that they capture in the lake. During the winter the nerpa swim under the frozen lake, creating breathing holes in the ice with their sharp claws. The nerpa can stay underwater for up to 70 minutes at a time. Crustaceans (aquatic animals with no backbone, jointed limbs, and a hard shell) have prospered in Lake Baikal. The most prevalent animal in Lake Baikal is the Baikal episclera crustacean.

It is exceedingly little, roughly the size of a grain of rice. It feeds on small algae and bacteria by filtering water through appendages around its mouth that seem like combs. The overall flow of water via the mouths of these little creatures is comparable to 10 times the flow of all the rivers that reach Lake Baikal each year. The labor of these microscopic creatures is attributed to keeping the lake so pristine. One of the most remarkable species in Lake Baikal is the golomyanka or oily fish. They are relatively tiny, around 10 inches (24 cm) in length, have no scales, and are translucent (Pathak & Mudgal, 2005). About 35 percent of their overall weight is made up of oil that is a highly pure type of vitamin A. Residents who have lived near the lake for many years report that the fish was used to cure numerous maladies, such as arthritis, and to calm wounds that would not heal.

Figure 4.6: Lakeview.

Source: https://media.istockphoto.com/photos/panoramasee-lake-in-stubaital-valley-in-tyrol-austria-picture-id1266619229?k=20&m=1266619229&s=612x612&w=0&h=jLQGc0IxgdlT0ZPMcHWQD94Kyr2viw-FZIslMdmD4W8=

4.4 BIOLOGICAL LIMNOLOGY

The purpose of biological limnology is to grasp the animals, plants, and microbes that inhabit lakes and rivers. The geology, physics, and chemistry of the lake or river impact the distribution patterns of these various animals. For instance, light is essential for plant development. Because water is so good at absorbing light, plants must either grow in shallow water along the coast or float on the surface of the ocean. Because the intensity of sunlight fluctuates with the seasons, plants frequently have a growing season in the spring, when light levels are at their greatest, and a mortality season in the fall, when light levels are at their lowest. Similar to humans, animals require oxygen dissolved in water to breathe. Coldwater contains more dissolved oxygen than warm water. Therefore, trout, which need a considerable degree of dissolved oxygen, are generally found in cold lakes and rivers. In contrast, bass requires less dissolved oxygen and is found in warmer lakes and lake surfaces. The introduction of alien species into lakes and rivers is one of the primary concerns of biological limnologists. People regularly introduce new species into lakes and rivers. In rare situations, non-native species develop quicker than native species and could take over the bulk of an area. In 1985, for instance, the zebra mussel was introduced into the Great Lakes of the United States by ballast water from the Caspian Sea in Asia. These mussels have become a big nuisance in the Great Lakes, blocking sewage systems and overgrowing docks and piers due to their fast reproduction (Owusu-Boateng et al., 2022). The fundamental purpose of Biological Limnology is to obtain a broad grasp of the concepts and literature surrounding the ecology and biology of aquatic animals. While species in lentic and lotic habitats will get the brunt of the attention, the wetland ecosystems will also get a thorough examination. One of the strategies used to explore species' interactions with biogeochemistry and with one another is synecology. The autecological history of the most major taxonomic groupings and hypotheses is covered in this section. We'll focus mostly on literature in this part. One taxonomic item is allocated to each student, and they are required to investigate and write or talk about the taxonomic history and ecological significance of that item.

Freshwater systems - lakes, marshes, rivers, and streams, have been crucial to the construction of civilizations throughout human history. From prehistoric times, civilizations have been formed dependent on their accessibility to water. Water bodies are crucial to humanity not just for drinking but also for transportation, agriculture, energy generation, industry, and waste disposal. Contaminated runoff from growing urban and agricultural zones, airborne pollution, and hydrologic disturbances such as the draining of wetlands are only some of the different components that continue to harm surface rivers. Determining which of these variables has the most important impact on the quality of a water body entails knowledge of the interaction of the water body with its watershed and how the various inputs affect its physical, chemical, and biological qualities. One of the key sciences necessary to comprehend these aquatic ecosystem interactions is termed limnology. Limnology is the study of fresh or saline water covering all aquatic ecosystems, including lakes, ponds, reservoirs, streams, rivers, and oceans. It is an interdisciplinary discipline that encompasses biology, chemistry, physics, and geology to investigate inland waterways as complex biological systems.

The formation of lakes is generated as a result of several natural and human activities, which are frequently connected resulting in lakes of comparable origin, and physical and biological properties. Lakes produced by glaciers: The gouging and scraping acts of glaciers have led to the production of lakes in the north temperate zones. Glaciers followed existing valleys, deepening and enlarging them. When the amount of ice melted, it generated heaps of moraine, which hampered valleys, damming streams and rivers. Ice-scour lakes: Where ice sheets travel over relatively flat surfaces of strongly jointed or fractured rock, hollow basins are produced and eventually filled with water. Lakes created by movements of the earth's crust: When a portion of the earth's surface dips in comparison to its surroundings or conversely the sides are elevated, a lake basin may be produced (Abubakar & Abubakar, 2013). The extent and depth of water will rely on the drainage from the surrounding land, which depends on the quantity of rainfall and the size of the region from whence water flows into the lake. Lakes generated by volcanoes: Many lakes are developed owing to volcanic eruptions classed as crater lakes. These lakes are fed exclusively by rainfall and runoff. They are discovered when a sequence of modest eruptions has taken place roughly next to one other, rather than one massive eruption.

In rare circumstances, if the volcano was enormous and erupted long ago, the bottom of the wide crater is relatively level and water gathers in the lowest regions. Volcanic activity also generates lakes when the ash or lava from an eruption blocks natural drainage. Lakes generated by rivers: When the flow of water is abruptly interrupted by a drop-in gradient, it pursues a longer route and travels at a slower speed. Sediments are eroded from the outside and deposited on the inside so that the channel becomes more convoluted. Sometimes the river breaches across a tiny isthmus between two subsequent bends. This may leave a loop of the river to one side of the new stream. The isolated part may hold enough water to form an "oxbow lake". Other lakes may be produced from hollows in the valley floor, filled by the river when it exceeds its banks during floods. The flat floor of a huge valley, which is continually inundated by the river, is known as a floodplain. In numerous regions of the world, coastal impoundment results in freshwater lagoons next to the sea. The freshwater lagoons are generated when a single strip of sand prevents the circulation of water from the river to the sea. Lakes are continually changing and follow the natural processes of transition and death (Ohtaka et al., 2011). Starting clean and deep, they progressively fill in owing to natural processes, increasing shallower until the wetland changes into a bog and finally ceases to exist. This is an extraordinarily slow procedure that takes thousands of years to conduct. This process dubbed natural eutrophication is inevitable. Human activity accelerates this process and affects the water quality within a few years. This is also regarded as cultural or anthropogenic eutrophication, due to which clear nutrient-poor water transforms into green, nutrient-rich water. The total quality and quantity of the water is depending on the physical features of the lake and its basin.

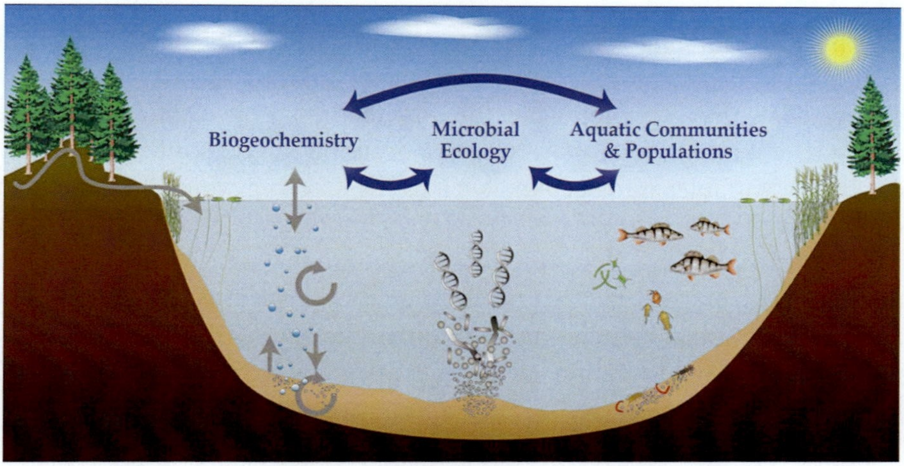

Figure 4.7: Image of limnology.

Source: https://www.ieg.uu.se/digitalAssets/429/c_429596-l_3-k_lake_limno_overview_060417-web.jpg

4.5 LENTIC AND LOTIC ECOSYSTEMS

Inland water bodies can be defined as either lotic (running water) or lentic (standing water). Lotic ecosystems include rivers, streams, and brooks, and lentic habitats include lakes, ponds, and marshes. The essential contrast between them is the ongoing flow of water in a lotic environment. Large deep basins frequently describe lentic ecosystems with little or no flow present inside the basin. Characteristic of lentic ecosystems is the development of vertical inequalities (vertical stratification) of several key features, which often display major seasonal variability as well. However, the important condition is the length of time a specific volume of water remains inside a specified region of aquatic habitat, a phrase explicitly tied to flow rates. Some big rivers with only a minor gradient have low rates of discharge and flow and vast floodplains with many connected basins of lentic waters.

4.6 PHYSICS AND CHEMICAL ENGINEERING FACTORS

The physical characteristics of the lake basin, such as its size, shape, and structure, define its morphology. Lakes with a large surface area and a shallow depth have poorer water quality than shallow depth lakes. The surface-to-volume ratio (S/V) is used to describe this. Lake Morphometry takes many variables into account, such as the lake's maximum and median depths, as well as the percentage of the lake's littoral area. A lake with a larger watershed area than its surface area receives more surface runoff with higher nutrient loads (Naik et al., 2012). Deep analyses, such as sediment depth measurement, strata volume, and shoreline characteristics, are frequently required in the investigation of fresh water's biological, chemical, and physical properties.

Some of the ecosystem components that require morphometric parameters to function properly are erosion, nutrient loading rates, chemical mass, heat content, thermal stability, biological productivity, and growth effectiveness. Management techniques such as effluent loading capacity and selective removal of undesirable biota components require a thorough understanding of morphometry and flow characteristics. Inputs minus losses govern the change in lake water storage, according to fundamental hydrology. Outgoing effluents, seepage, and evapotranspiration are equal to incoming precipitation, surface influents, and groundwater sources. Groundwater, drainage basins, climate, and seasonal and geographic variation all influence the inflows and outflows of lake basins. The hydrological cycle determines lake distribution to the catchment (watershed) area, which can be significantly altered by changes to surface water systems caused by humans.

Figure 4.8: Crater lakes are formed when the caldera fills with water.

Source: https://media.istockphoto.com/photos/summer-clouds-river-reeds-trees-picture-id996757422?k=20&m=996757422&s=612x612&w=0&h=9jOR-NgPrXS4hw6tgw_G8qofmxy23wnvGzmMJanK-ZE=

4.7 WATERSHED

Water must collect in the basin at the bottom of the catchment for a lake to form. The catchment area or drainage basin is the area from which water drains into the lake's basin. Water can enter a lake in a variety of ways, including directly from a stream, river, or other depression. The spatial extent and precipitation of the catchment have the final say in how much water enters the lakes. The watershed contributes to the body of water by allowing surface runoff to enter it. Any land or water that drains into a specific drainage channel is considered part of the watershed. Watersheds are divided into sections by a high line that separates areas that drain into different bodies of water, such as a small area of a few acres that drains into a small stream that drains into another area of a larger watershed, another area of another bigger watershed, another area of another bigger watershed, another area of another bigger watershed, and so forth. A tributary to a lake may have its watershed, which is a part of the lake's larger drainage area, but it is still considered a sub-watershed within the larger watershed (Mustapha, 2011). The watershed's size, topography, geology, land use, fertility of the soil, erodibility, and vegetation are all represented by the lake. Increasing the size of the watershed harms the quality of the lake water. Precipitation interacts with soil and leaches minerals before they reach a lake in a large watershed.

Groundwater flow is the primary source of nourishment for seepage lakes, which have small watersheds. Water from inflowing streams and rivers flows into drainage lakes. Land use has an impact on both the quantity and quality of water entering the lake. Increased surface water flow to the lake is facilitated by the high proportion of impervious surfaces in urban areas. This can lead to watershed erosion, allowing materials (such as minerals) to be carried along with the soil and negatively affecting water quality. An important part of a watershed is the type of soil and how vulnerable it is to erosion. The type and density of vegetation determine the extent of erosion. Erosion is less likely to occur in native, undisturbed vegetation compared to agricultural land. Another component of the watershed to be addressed is whether the lake is a closed (no outlet) or open (with outflow) system. In the closed system, all the water draining remains, not moving into a canal or river, making it more sensitive than in an open system. The upstream water quality has a crucial impact on the health of the lake habitat. Lakes in the upper end of the watershed may have a different water quality than the downstream. Depending on the physical and biological qualities of the upstream lakes, the downstream water quality may be altered either positively or adversely. Factors impacting watershed operations: Size: Both run-off volume and rate increased with an increase in watershed size.

Figure 4.9: A watershed.

Source: https://media.istockphoto.com/photos/underwater-beautiful-lake-between-the-green-banks-picture-id469266958?b=1&k=20&m=469266958&s=170667a&w=0&h=sGHVZmV2uVmPn24CIFeVCFZcbZJsL0_o6T6aK-0ZlViY=

However, both rate and volume per unit of watershed area decline as the area grows. The size or area of the watershed is a crucial component in determining the peak rate of run-off. Shape: Long and narrow watersheds have longer durations of concentration resulting in lower runoff rates than larger square watersheds of comparable area, which have some tributaries flowing into the main river. This sort of concentration also influences the amount of water seeping into the soil in the watershed. The longer time it takes to depart the watershed, the more is its seepage into the soil. Land-slope: Slope has considerable influence on land use. The velocity and amount of run-off depend on the slope of the ground. Bigger the slope, the greater the velocity of flow of run-off water. If velocity increases twice, energy and hence erosion likewise doubles. The degree of slope sets constraints on land use for annual crops, planting, and land restoration, depending on soil depth. Drainage pattern: The drainage pattern of a region relies on the course of the streams and their tributaries. Land slope, lithology, and structure impact the drainage pattern (M Ongaratto et al., 2018). In general, the larger the drainage roughness, the higher the conductivity. Finer drainage texture resulting in heavier soil type. Drainage patterns serve to help to find sensitive places needing varied sorts and degrees of soil conservation measures. Soil and geology: Soil and geology of the watershed also impact the quantity of water seeping into the earth. Soil type also impacts the quantity of silt that will be swept down into water collecting devices. Vegetative cover: The kind and quality of the vegetative cover of the watershed impact run-off, infiltration rates, erosion, sediment production, and evapotranspiration rate. Dense vegetation decreases erosion. Precipitation: The amount and kind of precipitation are critical components impacting the rate of run-off into the water body. Deluge dispersed widely throughout the year has a different result than a single peak seasonal rainfall.

4.8 WATER BUDGET IN LAKES

The water budget of lakes gives vital information on the availability of water in lakes at any point. The essential features of the water budget may be summarized as recharge and outflow. These may be further split as input from precipitation, surface runoff, and groundwater and outflow from rivers and evaporation. Hydrogeological, hydro morphological and hydrodynamic aspects impact the water budget in the lake basin. The water budget of lakes is governed by the circumstances of humidification in the area where the lakes and drainage basins are located. In humid locales, precipitation is higher than evaporation from the water surface, whereas, in arid regions, evaporation from the water surface is higher than precipitation. An intermediate zone may also emerge when both humid and dry zones may alternate (Medeiros et al., 2012). Each zone is distinguished by its structure and dynamics. Quantitatively, the water budget of lakes may be determined by following established procedures, the limits of which can be addressed using isotopic techniques. The usual way of measuring water budget comprises employing characteristics like surface inflow, lake precipitation, lake evaporation, surface outflow, subsurface outflow, and inflow and storage changes. The water budget of lakes may also be determined using environmental isotopes and artificial isotopes.

4.9 SEDIMENTATION

Because of particle deposition in the water, inorganic salt from neighboring farms and organic particles from dead plants and animals have collected on the lake bottom. These provide food for ocean-dwelling species and contribute to the chemistry of the water. Sediments in many lakes serve as significant decomposition sites. Big populations of microbes break down large organic molecules into smaller inorganic compounds in organic-rich sediments. As the oxygen in the sand depletes, bacteria in the water begin to use the remaining oxygen. When a lake stratifies, there is no hypolimnion mixing and, as a result, no oxygen replenishment to the bottom layers. The organisms die due to a shortage of oxygen in the surroundings. Only organisms that can withstand low oxygen levels in the hypolimnion will be able to survive in a huge lake. While some breakdown processes are slowed owing to a lack of oxygen in the sediments, other chemical reactions that occur primarily in the absence of oxygen take over. Anaerobic bacteria may live in the absence of oxygen, allowing them to break down enormous amounts of organic material in the mud and create methane gas (CH4) (made up of carbon and hydrogen CH4). Aerobic bacteria, which can only thrive in oxygenated settings, use methane as a source of energy. Other anaerobic bacteria that eat sulfates may make hydrogen sulfide (Matta & Uniyal, 2017). A layer of deoxygenated water at the bottom of many lakes is followed by a layer of oxygenated water. For the lake to be healthy, there must be a balance of aerobic and anaerobic activity occurs when you engage in aerobic activity while depriving your body of oxygen. Accelerated lake sedimentation has caused major problems, particularly in developing countries.

The operation and objectives of the water system could be significantly impacted as a result of undesirable sedimentation in the lake. A thorough examination of the catchment is required due to the following factors: Two critical steps in the process are measuring erosion in the catchment and locating sediment sources associated with specific land uses. Sediment storage areas must be planned ahead of time. The technology for measuring sediment storage is constantly evolving. These are the sedimentation rates that have been determined. The sedimentation of a lake tells us about the climate or anthropogenic changes that have occurred in its catchment area. The following factors have a significant impact on this process in lakes and reservoirs: Size and composition of nearby watersheds; high-relief areas contribute coarser particles, while low-gradient areas contribute finer particles.

The volume of a lake or reservoir influences sedimentation because it controls the waves in the system. The location of the lake or reservoir is determined by the wind direction. The movement of fine particles is influenced by thermal stratification and lake water density. Climate variables such as precipitation, evaporation, and temperature Rivers and streams transport sediment downstream to lakes and reservoirs. Fine sands (suspended/wash load), silt, and clay are among the sediments transported (depending on the water discharge, the bed slope, velocity, and turbulence). Most lakes are primarily sedimented by inorganic material, but in many shallow lakes, phytoplankton populations can contribute significant amounts of particulate matter to the lakebed (in areas of low relief, where rivers carry small loads). The flow rate and sediment concentration of the inflowing waters just upstream of the lake are the most common and accurate methods for determining the sediment load carried by streams to a lake. Another common technique is periodic lake bathymetric surveys. This can also be done with environmental isotopes. Less light reaches the lake's surface as a result of sedimentation.

Figure 4.10: A healthy lake with vegetation growing within.

Source: https://media.istockphoto.com/photos/summer-clouds-river-reeds-trees-picture-id996757316?k=20&m=996757316&s=612x612&w=0&h=W9BrueODkPXlDsmYjqJnw0NwklgzciQCfas9IMtpg0o=

4.10 LIGHT

Light is crucial to the ecology of lakes because it influences the potential rate of photosynthesis, which feeds dissolved oxygen and nutrients to the water. Solar radiation is the major source of heat, which impacts wind patterns and water movement in the lake basin. The large bulk of the energy that regulates the metabolism of lakes and streams is created directly from the solar energy consumed in photosynthesis. This energy is formed either inside the lake or stream (autochthonous) or inside the drainage basin, then transported in diverse ways to the lake or stream (allochthonous) Lake production depends on the successful conversion of solar energy to potential chemical energy, as well as the components that determine the efficiency of conversion. Seasons and depth influence light intensity. The greater the photosynthetic rate, the larger the light penetration depth. Photosynthetic organisms consist of phytoplankton (algae), attached algae (periphyton), and aquatic vascular plants (macrophytes). The number of suspended particles in water affects the rate at which light transmission falls with depth (Lund & McCullough, 2008). The highest depth at which algae and macrophytes may grow is restricted by the amount of light, which is defined as the depth at which available light falls to between 0.5 and 1 percent of the surface level. In a clear lake, the euphotic zone can run to depths of 20 meters or more, however, in the majority of current lakes, it is frequently no deeper than 3 to 5 meters. Human activities throughout the watershed have drastically changed the lake's transparency. Thermal structure and stratification patterns affect the feeding cycle, distribution of dissolved gases and biota, and behavioral adaptations of species. Consequently, the optical features of lakes and reservoirs are key physiological and behavior regulatory components for aquatic creatures.

4.11 DENSITY

Longer wavelengths, in particular, are absorbed near the lake's surface and transformed into heat. Water is the only chemical that is less dense as a solid than it is as a liquid (maximum dense at 4°C, less dense at both higher and lower temperatures). When the lake warms up, heat takes a long time to permeate down through the surface water. During the summer, temperature discrepancies between the upper and bottom layers become more noticeable. Deep lakes are frequently divided into three layers: epilimnion, metalimnion, and hypolimnion. The epilimnion is the upper layer of the ocean that is warm and well-mixed. The metalimnion, also known as the thermocline, is a layer of water underlying the epilimnion where the temperature swiftly falls with depth. The hypolimnion is the chilly water layer at the ocean's bottom. The density shift of the metalimnion works as a physical barrier, preventing epilimnion and hypolimnion from mixing. A thermocline is sometimes misunderstood with metalimnion; however, it is the plane or surface where the temperature drops the fastest in proportion to depth. As a result, the thermocline is the site of maximum temperature change in the metalimnion. Because of their depth and warmth, a few lakes stay continuously stratified, but countless tiny lakes are never stratified for more than a few hours (Leipe et al., 2014). Most lakes, on the other hand, experience seasonal stratification and mixing, which has a crucial function in their ecology. Thermal stratification is influenced by solar radiation and wind, as well as heat exchange. The size, wind exposure, and amount of water inflow of a lake all have a part in establishing the sort of circulation. A lake's circulation patterns can be categorized based on its degree and frequency.

Circulation pattern in lakes: Amictic: never combine since they are continually frozen. Meromictic: mingle only partly, the deeper strata never mix either because of high water density induced by dissolved chemicals or because the lake is sheltered from wind influences. Holomictic: fuse entirely. Oligomictic: do not mix every year as they are huge and have higher heat storage capacity, the mixing relies on unique climatic conditions. Monomictic: combine just once per year, either in winter or summer. Dimictic: mix twice a year and are the most frequent lake in temperate latitudes. Polymictic: mixing regularly, they are shallow tropical lakes with high wind exposure.

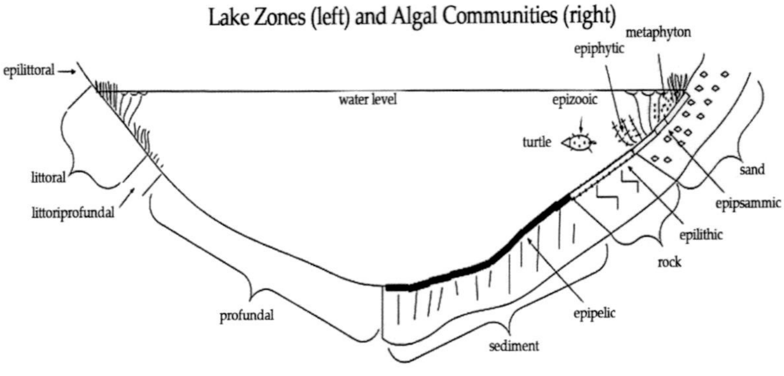

Figure 4.11: Density.

Source: By Pdietry - Own work, CC BY-SA 4.0, https://commons.wikimedia.org/w/index.php?curid=97678384

CHAPTER 5

POLAR LIMNOLOGY

CONTENTS

5.1 INTRODUCTION

Lakes, rivers, and streams may be found in both the Arctic and the Antarctic. These bodies of water can be found in both Polar Regions. The food web structures in these ecosystems are simpler than those found in the latitudes that are classified as temperate, even though these ecosystems offer a greater diversity of aquatic habitats for the populations of biological organisms that live there. The ecosystems that are found in high-latitude water habitats and the biota that inhabit them make for excellent models for a wide variety of scientific fields. Some of these fields include ecology, microbiology, paleoclimatology, astrobiology, and biogeochemistry, to name just a few (Larsen & Alp, 2015). Other fields include astrobiology, which studies the evolution of life in space, and astrobiology, which studies the evolution of life on earth. Native Americans who make their homes in northern regions are wholly reliant on the hunting and fishing possibilities offered by the waterways in this region.

Arctic populations are dependent on ice sheets for more than simply their water supply; they also utilize them for energy, transportation, and mining. Ice sheets provide all of these essential services. Not only do these streams have breathtaking limnological characteristics, but they also act as sentinels for global warming and as producers of greenhouse gases. In addition, the immense rivers of the Arctic are a vital supply of fresh water for the oceans that encircle our globe; hence, these waterways are a very important component.

Polar areas are undergoing climate change at a more rapid rate than the rest of the planet, which has spurred curiosity in several peculiarities of the region, including the region's distinctive internal rivers. Despite the many similarities that exist between Antarctica and the Arctic, the water ecosystems of the two locations couldn't be more different from one another. High latitude regions in both the south and the north are characterized by low temperatures, severe snow and ice effects, limited annual solar energy inputs, and significant seasonal shifts in the light and temperature regimes. This is the case because high latitude regions receive less direct sunlight.

Both climate change and colonization have the potential to have a substantial influence on the region of the Arctic. This is because the Arctic is essentially an extension of continental landmasses to the north. Terrestrial vegetation is frequent in catchments that are located in the Arctic, but in catchments that are located in the Antarctic, it is unusual to find it. When compared to Antarctica, the quantity of allochthonous organic carbon that is found in the lakes of the Arctic is a substantial amount smaller than that which is found in the lakes of Antarctica (Kumar, 2022). The animal, plant, and microbial communities that can be found in Arctic waters are more diverse than those that can be found in Antarctic seas because of the proximity of Arctic waters to the northern temperate zone.

Figure 5.1: Wet Polar Bear Photos.

Source:https://thumbs.dreamstime.com/b/wet-polar-bear-emerges-coat-swimming-ice-floe-to-ice-floe-cold-arctic-ocean-taken-svalbard-65997413.jpg

Additionally, the food webs that can be found in Arctic waters are more intricate than those that can be found in the Antarctic seas. There are no fish in any of Antarctica's lakes or streams, and some of the lakes that are located in the south polar area do not even have zooplankton in them. Although it is possible to find insects in Arctic lakes and rivers up to the Arctic Circle, to this day only two types of insects have been confirmed to live in Antarctica. This is even though it is possible to find insects in Arctic lakes and rivers up to the Arctic Circle. Benthic ecosystems may be found in the waters of both of these locations, and they are characterized by the predominance of microbial mats that are dominated by cyanobacteria.

On the other hand, it is possible to find aquatic mosses in the streams and lakes of Antarctica. This discovery might be made in the future. By using comparative limnology of the Polar Regions, researchers can investigate questions such as the factors that influence the biogeography of aquatic plants, animals, and microorganisms; the limiting factors for biological production; the causes and consequences of food-web complexity; and the responses of aquatic ecosystems to environmental change as a result of these similarities and differences.

5.2 THE ARCTIC LAKES

Limnologists have known for a very long time that lakes that are found at higher latitudes will have a unique set of properties, and they have made this prediction. Francois-Alfonse Forel, a pioneer in the field of limnology, predicted that the water temperature of Polar Lakes would never rise above 4 degrees Celsius due to the short summers and low light angles that occur at high latitudes. He reasoned that this would be the case because high latitudes have shorter summers. According to the taxonomy of polar lakes developed by G. Evelyn Hutchinson, "cold monomictic" lakes can exist at both high latitudes and considerable heights. In addition, each of these lakes possesses a monomictic composition, which means that they are home to a single kind of living thing. It is also likely that some of the low-arctic lakes may not properly circulate their salinity gradients, whereas other lakes do. However, this is not the case with all of the lakes (Knight et al., 2015). Researchers in Alaska, the United States, Greenland, and Antarctica started conducting studies on the temperature patterns of polar lakes in the 1950s and 1960s.

These temperatures were measured in degrees Celsius. At first, the study that was done on the aquatic ecosystems of the Arctic was restricted to just a few site visits, which were referred to as "descriptive." There is a fair possibility that Juday found a sample of cladocerans and zooplankton from the Canadian Arctic Expedition of 1913–1918 in Point Barrow in the year 1882. This was the year when the First International Polar Year was being observed. During the decade of the 1950s, researchers in the Arctic and Antarctic carried out a substantial number of observations in lakes, most of which were restricted to the summer months. Outstanding work was done by Ulrik Ren in his study of the freshwater biology of Disko Island, which is located in Greenland.

The vast majority of the most significant discoveries were made in the 1960s and 1970s, and they were made by investigating several processes either concurrently or sequentially over lengthy periods. These decades saw the bulk of the most significant advances in science. Research on the connection between human productivity and well-being was carried out by the International Biological Programme (IBP) beginning in the year 1970. In addition to the various ponds and lakes that were spread out over the Arctic, there were also lakes and other bodies of water to be found in the cities of Barrow, Alaska, and Char Lake, in the country of Canada. Three to four years were spent by researchers at these institutions looking at every facet of limnology, from the tiniest bacterium to the largest fish. Scientists were able to make important achievements that went beyond the boundaries of descriptive limnology by researching the factors that induce and regulate the spread of Hux in vast aquatic systems. This was made feasible as a consequence of a comprehensive inquiry that was based on questions and was carried out methodically.

The research that was carried out in Barrow covered both terrestrial and aquatic components, whereas the work that was carried out at Char Lake concentrated on the lake itself and included comparison studies with Lake Marietta, which had become eutrophic as a result of the importation of sewage. Alaska served as the location for both of these initiatives. Due to the investigator effect, the pond in question required to be connected to an overhead railroad to make room for the 29 researchers who were going to sample it. The successes of the International Biological Program (IBP) provided the incentive for the United States to launch an initiative of integrated ecological research at a total of 26 sites, the bulk of which are situated within the United States (Knott et al., 2002). This program is being carried out in the United States. Both at Alaska's Toolik Lake and Antarctica's McMurdo Dry Valleys, researchers are hard at work collecting data and conducting experiments.

It wasn't until 1975 that researchers discovered Toolik Lake in the McMurdo Dry Valley, and it wasn't until the late 1950s that they discovered McMurdo Dry Valley Lake. Both lakes are located in the McMurdo Dry Valley. Even though it is anticipated that each LTER project would continue for decades, create publications, provide jobs for graduate students, and collect data that can be accessible over the internet, the projects are only examined once every six years.

Figure 5.2: Arctic Ocean.

Source:https://media.istockphoto.com/photos/ice-floe-picture-id454991203?k=20&m=454991203&s=612x612&w=0&h=YhGHkNk5vHYQQZV4Oyg1j6FKWSJn5CAt3pj_wb_bT9k=

For the foreseeable future, the Arctic Long-Term Ecological Research project's research on the long-term prognosis of climate change will concentrate on lakes, streams, and tundra. In a larger sense, the objective of the McMurdo Long Term Ecological Research project is to get a greater knowledge of how the structure and function of dry valley ecosystems are influenced by physical and biological limits, in addition to the movement of material. The fact that the IBP and LTER are models of whole-system and synthetic approaches to the study of limnology must be brought to the notice of as many people as possible. An approach that is taken over a protracted period paves the way for the collection of extensive climatic data, data sets that span decades, the study of the entire system, and several integrated investigations into the major physical, chemical, and biological processes that occur in particular areas.

These can all be accomplished by taking the approach. This kind of research is essential; however, it is also necessary to do more extensive sampling of Polar Regions to gain a better understanding of the impact that a variety of geological and climatic elements have on the Polar Regions. This can be accomplished by conducting research in both Antarctica and the

Arctic. The islands of Signy and Livingston, as well as the Vest fold Hills and the Schirmacher Oasis, are both significant sources of limnological data for the maritime zone of Antarctica. Some of the lakes in the circumpolar Arctic area that has been studied by researchers include those located in Alaska, Canada, northern Finland, and a significant amount of Greenland, Svalbard, and Siberia (Kazanci et al., 2004). They are taking an ever-increasing interest in moving waterways, like the ephemeral streams that meander through the McMurdo Dry Valleys and the huge Arctic rivers that flow through the lake-rich Hood plain.

Specifically, they are looking for evidence of life in these rivers and streams. Other themes, such as high-latitude limnology, the ancient limnology of Ellesmere Island's north, the limnology of the Vest fold Hills, and the influence that climate change has had on northern freshwaters, have also been explored in special journal editions. All of the freshwaters, rivers, and wetlands of this region were dissected in great depth within the arctic limnology articles that were written on the Schirmacher Oasis and the McMurdo Dry Valleys. A book that offers an overview of the subglacial aquatic ecosystems that are found beneath the ice sheet that surrounds Antarctica has been recently released by the National Academy of Sciences of the United States of America. In 2006, Bergstrom disseminated his results to the scientific community on the present condition of Antarctic ecosystems.

5.3 CATEGORIZATION OF ORGANISMS IN THE FIELD OF LIMNOLOGY

The area of Antarctica that is south of the Polar Front Zone and is commonly referred to as the Antarctic Convergence is home to a wide range of subglacial aquatic habitats. These ecosystems include several lakes, ponds, and streams that are located along the coast. These kinds of environments may be discovered throughout the continent. An example of a lake and river ecosystem is the southern boundary of discontinuous permafrost in the eastern Canadian Arctic, which currently reaches the southern end of Hudson Bay in the easternmost section of the country's Arctic. This region is located in the easternmost part of the country's Arctic.

The part of the Arctic that is furthest to the east is where you will find this particular habitat. Each of these groups is influenced in some way by the changes that have taken place in the world around them. It is possible to discover the forest-tundra Eco zone in these northern nations. This Eco zone is often referred to as the Sub-Arctic or the Low Arctic. After this Eco zone comes the genuine tundra, then the shrub tundra, and finally the polar desert of the high Arctic. Lakes with different levels of acidity and alkalinity may be found in the Polar Regions. These lakes can be found in both freshwater and saltwater forms (Jeong et al., 2008). Certain lakes do not completely ice over throughout the wintertime, and those can be found. The fact that these ecosystems' summer thermal regimes can range from fully mixed to thermally stratified throughout a temperature range of 40 degrees Celsius is a reflection of the complexity of these ecosystems. During the summer, these ecosystems' temperatures can be anywhere from 20 to 40 degrees Celsius.

Figure 5.3: Freshwater ecosystems in the Arctic.

Source:https://storage.googleapis.com/oceanwide_web/media-dynamic/cache/widen_1600/media/default/0001/05/650d984c9040e03f94d10ae6e2662292fa1b4fb2.jpeg

Conditions were maintained throughout the summer. The variation in shape is intricately connected to the chemical circumstances, which can range from oxygen super saturation to anoxia, usually within the same lake but at different depths. The water bodies that are generated as a result of the melting of the permafrost are keeping a vast diversity of animals, benthic ecosystems, aquatic plants and plankton, as well as migrating birds, alive. Melting permafrost also affects the environment in other ways. In the tundra, these different bodies of water serve as distinct bio-clusters. According to the findings of this research, waterbodies of this kind have been investigated in the delta of the Mackenzie River, for example, and the total number of water bodies in the delta of the Yukon River is predicted to be greater than 2,000,000. Both of these deltas are located in the Northwest Territories (Islam & Chowdhury, 2013). On the other hand, the bulk of thaw lakes are no deeper than a few meters at most, but the permafrost lakes that are the deepest can reach a depth of twenty meters.

When the temperatures go very low, shallow rock-basin ponds are widespread not just in the Arctic but also in the Antarctic. This type of body of water may be found in both of these regions. Lakes that are heated by the sun and do not become frozen throughout the year are something that can only be found in the Arctic and the Antarctic. Epishelf lakes are freshwater lakes that are situated on the landward side of ice shelves that lay on top of colder and thicker oceans. These lakes go by the name epishelf lakes, and they have the same name as their location. Milne Fjord and Beaver Lake are two places in the Arctic that particularly stand out to visitors. The ice sheet that covers Antarctica is home to a large network of subglacial aquatic habitats, one of which is Lake Vostok, which contains waters that are massive, incredibly deep, and shrouded in mystery.

Lake Vostok is one of the subglacial aquatic ecosystems that can be found beneath the ice sheet. Some rivers and streams run along the boundaries of Antarctica, however, they are seasonal rivers and streams, meaning that they only flow for a few months out of the year. Only during those few months of the year does the vegetation that lives there bloom. In the Arctic, which is home to massive catchments as well as some of the world's largest rivers that flow into the Arctic seas, surface waters play a significantly more vital role (Ishida et al., 2020). This is because the Arctic is home to some of the world's largest rivers. Because of the activities of both animals and humans, a great number of arctic lakes that were formerly extremely low in nutrient density or ultra-oligotrophic have undergone a large rise in the amount of production taking place in those lakes.

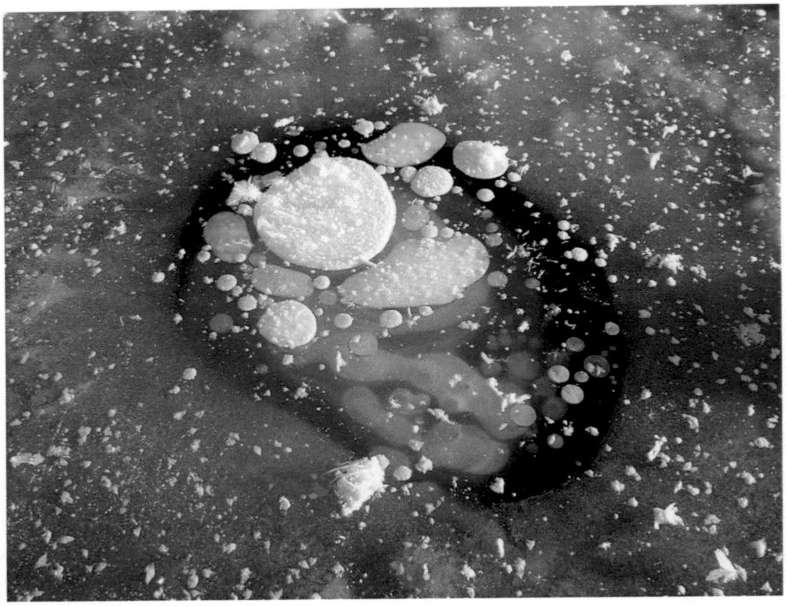

Figure 5.4: Methane Bubbles from Arctic Lakes Measured from Space as Atmosphere.

Source: https://commons.wikimedia.org/wiki/File:Frozen_Methane_Bubbles_%2811927906233%29.jpg

5.4 MANAGING THE FACTORS THAT AFFECT THE CREATION OF BIOLOGICAL MATERIAL

In ecosystems characterized by high latitudes and freshwater, what factors or factors may be responsible for limiting the amount of biological production? Not only is it a significant topic for people who study polar limnology, but it may also provide insight into the factors that are influencing aquatic output in other parts of the world. This knowledge is necessary for properly estimating how inland water ecosystems will react to the significant physical, chemical, and biological changes that will almost surely precede future climate change. These shifts are an almost definite thing to take place.

5.5 THE ACCESSIBILITY OF WATER SOURCES

The yearly cycle of freezing and thawing creates significant restrictions on the water supplies that are accessible in the polar areas, which are critically necessary for marine life to survive. The length of time available each year for biological activity is limited to a few days to a few weeks due to factors such as ephemeral streams in Antarctica and melt water lakes on polar ice shelves, for example. Even the tiniest of ponds may have a thin layer of water covering the inhabitants of the benthic layer if the snow and ice cover is substantial enough to prevent the lakes from freezing over during the winter. This is because lakes freeze over when the water temperature drops below 32 degrees Fahrenheit (Hodgson et al., 2001). Because they are nourished by water that comes from lower latitudes, the huge rivers of the Arctic continue to flow, albeit at a more leisurely speed, even in the depths of winter. This is because the rivers receive their water supply from further south.

As a consequence of this, it is projected that several polar aquatic ecosystems will be microbiologically active throughout the entirety of the year, with large seasonal fluctuations being produced by variables that are not connected to the quantity of water that is present in the environment (or in addition to this factor). The prolonged, 24-hour exposure of snow and ice to solar radiation during the summer and the low-temperature evaporative loss both contribute to an increase in the amount of melt water that is available for use by aquatic ecosystems (Barbosa et al., 2012). This increase can be attributed to the fact that snow and ice are more likely to melt during the summer. The glaciers and snowpack that melt in the Arctic are what feed the freshwater ecosystems that are found there, and major lakes are generated when there is considerable snowfall in the spring in the Hooding of the Hood plain.

5.5.1 Irradiance

Solar energy does not reach the poles until it has travelled to a lower latitude, which is made even more problematic by the fact that snow and ice have the effect of decreasing the amount of irradiance that reaches the ocean. This has a substantial effect on photosynthesis, which is halted when the sun goes down and starts back up again when the light rises. Yearly productivity in aquatic environments in the Arctic and Antarctic is decreased as a direct consequence of this. However, it does not appear that the quantity of irradiance that is absorbed by the water below the surface is the major element that influences the significant disparities that exist between lakes in terms of the daily primary production of phytoplankton throughout the summer. Phytoplankton biomass and photosynthesis may reach their maximum levels beneath the ice in the early spring, even though the amount of irradiation in the upper water column is growing (Helm, 2017). However, they may subsequently experience a significant reduction during the summer. This is most likely because there is less food available. Even though the incident irradiances are equivalent, the phytoplankton biomass densities and photosynthetic rates in nearby lakes in the Canadian High Arctic, such as Marietta and Char, as well as Fryxell and Vanda in the McMurdo Dry Valleys, are not the same.

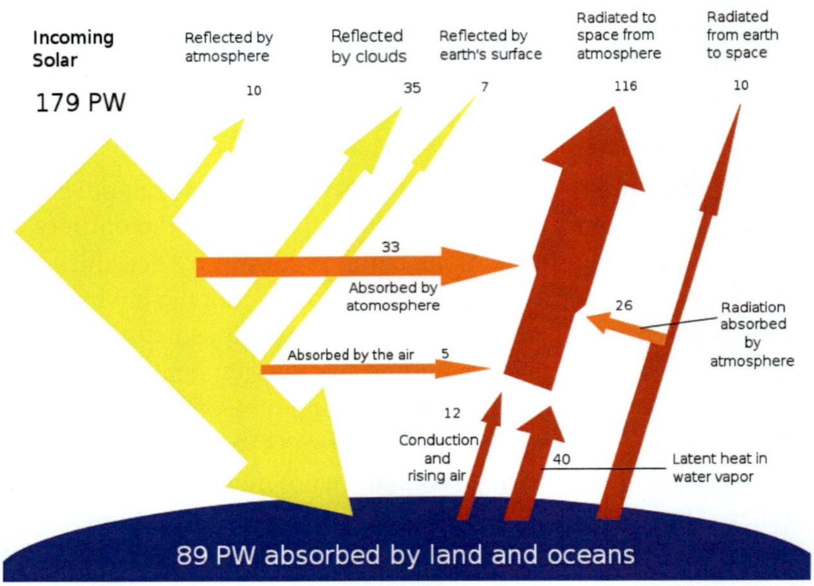

Figure 5.5: Solar irradiation.

Source: By User A1 at English Wikipedia, CC BY-SA 3.0, https://commons.wikimedia.org/w/index.php?curid=4393104

5.5.2 The Frigid Weather That Is Outside

Despite the widespread belief, there are a few aquatic ecosystems in the Arctic that maintain good summer temperatures, and some of these habitats even maintain temperate temperatures into the winter. Diurnal heating can occur in northern lakes if there is a large concentration of light-absorbing dissolved organic matter and particles. This can cause the lakes to reach temperatures that are higher than 15 degrees Celsius. Lakes in the south can reach temperatures that are lower than 10 degrees Celsius, whereas shallow lakes in the north can reach temperatures that are higher than 10 degrees

Celsius. Despite being located at a latitude of 69 degrees north, the mouth of the Mackenzie River may see temperatures as high as 17 degrees Celsius in the summer (Balarabe, 2006).

Radiation from the sun can heat stratified lakes, which are lakes that are always covered in ice, for decades or even millennia at a time. In both of the polar areas, there are examples of lakes that are meromictic and are heated by the sun. These lakes have maximum temperatures that are discernibly higher than the temperatures of the air during the summer, and those temperatures may be up to 70 degrees Celsius higher than the temperatures during the winter. The bulk of polar lakes has liquid water temperatures of 0 degrees Celsius beneath the ice, whereas subglacial Antarctic lakes that are topped by kilometers of ice have temperatures of 2 degrees Celsius. This is in contrast to the intense cold of the surrounding atmosphere. These temperatures create thermal conditions that are suitable for the functioning of biological organisms.

The average temperature of the water in polar aquatic ecosystems is very close to 0 degrees Celsius for most of the year. This is because the Polar Regions experience extremely cold winters. It would appear that the bulk of the creatures that may be found in these habitats are cold-tolerant rather than cold-adapted; the extremely low temperatures may limit their metabolism and slow down their pace of growth. This might be the case. There is a chance that certain lakes that are located at high latitudes have vast stocks of aquatic biota even though temperatures below freezing impede photosynthesis and other physiological processes. On the other hand, it is not certain that lakes with higher temperatures would have higher concentrations of phytoplankton or higher production rates.

5.5.3 Obtainability of a Wide Variety of Nutrients

Because of the interaction between the characteristics of the arctic lake and the watershed, the nutrient supply rates that were available for biological production were insufficient, particularly for the plankton groups. This was especially the case for the larger plankton groups. The low temperature, low moisture, and freezing conditions limit the microbial activity in the soil. Additionally, these circumstances delay all of the geochemical processes, including the reactions that occur during weathering in the soil. Because of this, there is a reduction in the amount of nitrogen that is released into groundwater and surface runoff, both of which are reduced How when the temperature is exceptionally low (Bassi et al., 2014). It is difficult for plants to flourish in the severe climate of the Arctic because it lowers the quantity of root biomass, associated microbes, and organic matter. This makes it tough for plants to grow. All of these parts are required to set in motion the processes of weathering.

The low temperatures that are typical of the oceans that surround the Arctic and Antarctic have an additional influence on the rates of nutrient recycling. Most of the year, ice cover protects polar waters from the wind-driven mixing that would otherwise take place in those areas. Because of this, the vertical movement of nutrients from the waters at the bottom to the zone exactly beneath the ice, which is the zone with the maximum concentration of solar energy for primary production, is greatly inhibited. This zone is the zone where the ice is directly on top of the water. In addition to this, it creates stratified habitats, which allow inflowing streams to avoid the primary body of the lake without bringing their nutrients with them, and it encourages the deposition of particles to reduce the levels of nutrients in the water.

Both of these things help to reduce the number of nutrients in the water. Intake of food, in addition to light and temperature, has been reported to have a significant role in the development of phytoplankton in polar lakes by several distinct lines of study, which suggests that the importance of this factor cannot be overstated. First, even though the irradiance and temperature regimes of different streams within the same region are comparable, there are significant differences in the primary production of these streams due to the various nutritional qualities of the water. This is the case even though the irradiance and temperature regimes of these streams are similar. Second, stratified oceans contain the highest rates of primary production and the greatest stocks of phytoplankton that are constantly reproducing (Benito et al., 2020). This is because stratified oceans contain the biggest standing stocks of phytoplankton deep inside the water column, where light access is limited but nutrient supply rates are higher. Stratified seas also have the highest nutrient supply rates.

Third, ponds that have been influenced by penguins on Ross Island and Cierva Point in Antarctica, as well as Marietta Lake in the high Arctic, which was enriched with human sewage, have significantly higher algal biomass stocks and production rates. This was found to be the case in all three of these locations. This is the case for anybody of water in either polar area that has been supplemented with nutrients derived from either natural or human sources. According to the findings of nutrient bioassays carried out in Toolik

Lake, lakes in northern Sweden, Ward Hunt Lake in the Canadian Arctic, and McMurdo Dry Valley lakes, plankton assemblages in high-latitude lakes have a significant reaction to the addition of nitrogen. These lakes are located in Antarctica and the Canadian Arctic.

Even though nutrients have the potential to be a significant productivity limitation due to the effects of the Liebig type on ultimate yield and biomass standing stocks, secondary Blackman-type effects on production rates per unit biomass, and therefore specific growth rates, can also occur. This is the case even though nutrients have the potential to be a substantial productivity limitation. Cold temperatures affect the maximum, light-saturated, and to a lesser extent, light-limited photosynthetic rates. However, low irradiances beneath the ice may drop primary production even more. Nutritional stress, which slows the growth of cellular components such as light-harvesting proteins and photosynthetic enzymes, may be a reaction to low light or low temperature and may magnify the effects of either of these environmental factors. Nutritional stress may occur in response to either of these environmental factors.

5.6 ARCTIC LAKES AS HOME TO COMPLEX FOOD WEBS

In the lakes of the Arctic, there is no well-established food chain. On the other hand, food chains can be traced from lakes in the low Arctic that are rich in zooplankton and fish to lakes in the high Arctic and Antarctic that are extremely cold and have ecosystems that are dominated by Hagellates, ciliates, and rotifers. These lakes are at opposite ends of the planet. The biogeography of the lake, which is influenced by the lake's trophic status, shapes the composition as well as the variety of the food webs that are found inside the lake. Although many Antarctic lakes are devoid of crustaceans and only a few species have made it to the continent, some of the continent's lakes likely have crustacean zooplankton (Hassan et al., 2018). This is even though many Antarctic lakes are completely devoid of crustaceans. Because of its oligotrophic qualities, Toolik Lake in Alaska is a great example of a low Arctic lake. This is because the lake is located in the state of Alaska. The bulk of the components that make up the planktonic food chain is small photosynthetic Hagellates and bacterial cells.

The dissolved chemicals in the watershed provide these creatures with the organic ingredients that they require for survival. There have been reports of micro-Hagelites in rare cases that are photosynthetic and feed on bacteria. Crabs and rotifers, which come in a wide variety of species, use these organisms as a source of food. As a direct result of this, three of the five different species of fish rely on zooplankton as their primary source of sustenance. Nevertheless, even though this food chain appears to adhere to a predetermined format, there are substantial variations from this pattern. Nanoplankton is responsible for the vast bulk of the production, which is to say that they are accountable for ninety percent of it. Second, there are not enough zooplankton in the water because there is not enough primary production. Zooplankton grazes on algae to lower their numbers, but there are too few of them. Third, there is a lack of zooplankton, which is detrimental to the growth of the fish since it prevents them from feeding.

As a direct result of this, zooplankton is the very last link in the food chain that begins with phytoplankton. Studies that made use of stable isotopes concluded that the benthos that can be discovered in Toolik Lake acts as the principal source of food for the fish that are present in that lake. In this particular example of a benthic food chain, the diatoms that are present on rocks and sediments are the source of energy for snails. Snails, in turn, pass this energy on to chironomid larvae through the detritus that they ingest. Chironomids, which are afterward consumed by lake trout, are finally digested by small fish, which are again given to snails by lake trout. This cycle continues until the food chain is complete. Rigler found a similar benthic-based feeding network in Char Lake, which only featured Arctic char and one type of copepod as its principal food supply for fish. The lake was named after the discoverer, Rigler. You may discover char that subsist on benthic crustaceans in some of the lakes that are located the farthest north in North America, specifically Lake A, which is situated at a latitude of 83 degrees north (Gothwal, 2021). Bacteria, odorless Hagellates, and algae predominate as the dominant species in the planktonic food webs of Antarctic lakes. These waters are completely devoid of any mammalian life.

The F-shaped body plan that is typical of all other living organisms is absent only in metazoans. There is a sizeable quantity of photosynthetic phytoplankton as well as mixotrophs in this particular sample. They derive their power from the sun and nourish themselves by consuming the cells of other microorganisms. The microbial loop, also known as the carbon-energy cycle in nanoplankton, is an essential activity that takes place in the lakes of both the Antarctic and the Arctic. Both of these regions include lakes. It is essential to take into account not only the primary production of planktonic organisms and the synthesis of dissolved organic carbon, but also the uptake of DOC by bacteria and the transfer of DOC to higher trophic levels via nanoflagellates, ciliates, and rotifers. This is because all of these processes have an impact on the total amount of DOC in an ecosystem. Lake Fryxell, which is found in the McMurdo Dry Valleys, is one of numerous lakes found in both the Arctic and Antarctic that include picoplankton and thin oscillatorians.

Other lakes like this may be found in both regions. You'll discover an especially high number of solitary planktonic crustaceans in places like the Vest fold Hills that are devoid of ice cover. In marine-derived salty lakes, you may find both *Daphniopsis studeri* and *Paralabidocera antarctica*. However, *Daphniopsis studeri* favors freshwater or slightly brackish lakes, but *Paralabidocera antarctica* lives in marine-derived salty lakes. There is a chance that the McMurdo Dry Valleys can be located both within the Antarctic Peninsula and to the south of it. Both of these locations are conceivable. Planktonic crustaceans are difficult to find in the benthos, even though there are a few copepod nauplii and two different kinds of rotifers present. It has been shown that the trophic status of lakes has an effect on the composition as well as the variety of food webs in both the Polar Regions (García-Girón et al., 2020). This relationship has been investigated to a significant degree. Ultra-oligotrophic describes the condition of Type I lakes, which are characterized by extremely low amounts of primary production.

According to this categorization, the predominant population of these lakes consists of algae, bacteria, nanoflagellates, and ciliates. Rotifers are a kind of microzooplankton that are particularly abundant in Type II lakes, which are distinguished by having significantly increased rates of productivity. Lakes that are located at the furthest north of the Arctic include examples such as Ward Hunt Lake and the McMurdo Dry Valley. Other examples of lakes in this region include the McMurdo Dry Valley. There is a chance that the copepod population will rise as a result of the greater production in Type III lakes like Char Lake.

In lakes of variety IV, which make up the vast majority of the world's lakes, it is possible to find Cladocera and Copepods. Even though the ponds in Barrow have been frozen over and do not have any fish in them, they are still considered to be a part of this group. However, the reason (or reasons) behind why fish could be found in Type I and Type II lakes that are located in the far north are not known. The char eats chironomid larvae, which are the stage of the benthic food chain that is the most common and prolific. The chironomid larvae are ingested by the char. On the other hand, it is widely believed that not a single one of Antarctica's lakes is home to any form of aquatic life, of any type.

5.7 A GENERAL WARMING TREND OBSERVED IN THE POLAR LAKES OF THE PLANET

The Polar Regions are vulnerable to a wide variety of threats, the likes of which are increased pollution imports, increased exposure to UV radiation, and climate change, all of which are expected to become more severe in the foreseeable future. One example is the Arctic, which is vulnerable to pollution imports. It is anticipated that the temperatures in the high latitudes will rise the most considerably during this century as a result of a wide variety of feedback processes that will accelerate the warming that is already occurring in these areas (Gagan et al., 2018). This is one of the most important factors that contribute to the fact that this is the case. The melting of ice and snow improves the amount of solar energy that can be used for heating because it increases the amount of water vapor that can be sustained in heated air. Both warm air and water vapor contribute significantly to the greenhouse effect.

In addition to that, it is projected that there will be major alterations in the distribution of the rain that falls. It is expected that there will be an increase in the amount of rainfall that occurs in the Arctic and Antarctic, as well as an increase in the frequency of rainfall, particularly at the highest latitudes, where it is unusual for there to be rainfall. This is one of the effects that is expected to result from climate change. Lakes that are found at higher elevations have already begun to experience the negative consequences of climate change. In extreme circumstances, the disappearance of permanent ice cover, an increase in water temperatures, the stratification of the water column, and other changes in the water balance can result in lakes and wetlands being fully drained or dried out entirely.

This is due to the domino effect that these elements have on one another, which causes the chain reaction. Several polar aquatic ecosystems have experienced significant shifts in the values of limnological parameters as a result of relatively minor changes that climate change has brought about in the physical, chemical, or biological environment. These changes have been brought about as a result of climate change (Eaton & Kardos, 2015). These threshold effects can bring about rapid and unexpected modifications in the structure and function of an ecosystem, in contrast to shifts that are gradual and may be predicted in advance.

Figure 5.6: Large Male Polar Bear in Canadian Arctic.

Source:https://thumbs.dreamstime.com/b/large-male-polar-bear-canadian-arctic-standing-frozen-lake-219426495.jpg

5.8 RESTRICTIONS PUT UPON US BY OUR NATURAL ENVIRONMENT

The existence of still water and the relatively untouched nature of the lake basin are the two aspects that are the most significant. According to the findings of a recent study, the melting of permafrost soil in Siberia has resulted in the drying up and disappearance of several lakes, which has, in turn, led to the extinction of a large number of different species. It is the first time in millennia that ponds in the High Arctic have evaporated due to a change in the equilibrium between precipitation and evaporation. This has been triggered by a shift in the equilibrium between precipitation and evaporation. There has been an increase in the size of habitats that are located in shallow water due to the melting of permafrost in specific regions over the past fifty years (Degefu et al., 2014). These regions include places that were formerly frozen solid. Research that looked at patterns over a longer period concluded that climate change has a wide variety of consequences on the lakes and rivers in the Hood Lain region, including the Mackenzie River.

Lakes found at higher elevations are less protected from the elements and have a larger potential to evaporate over time as a result.

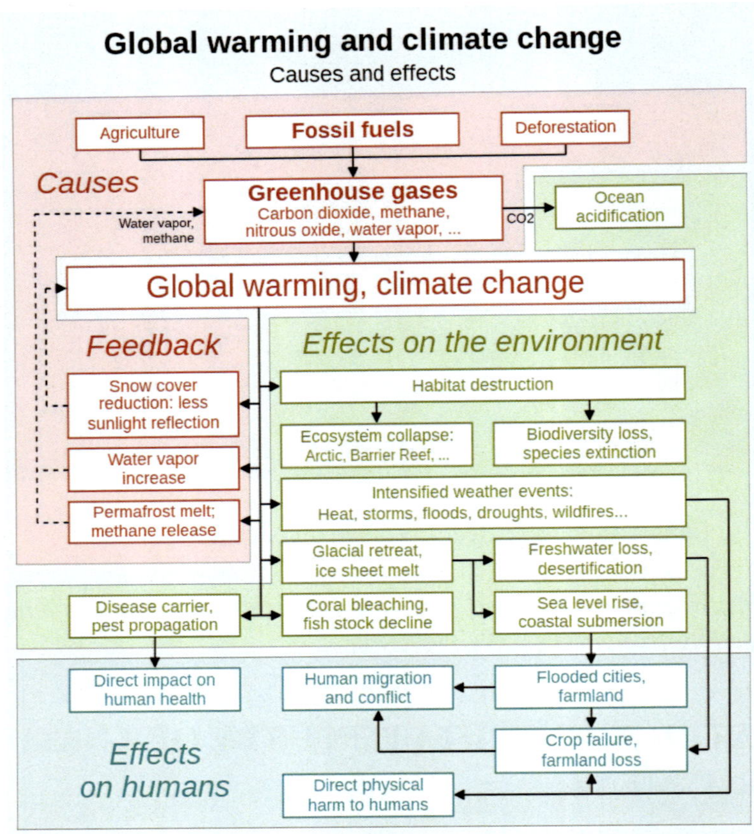

Figure 5.7: Pressure on the environment.

Source: By RCraig09 - Own work Also, successive versions were developed through collaboration with users Efbrazil, Femkemilene, and J. Johnson on en.wikipedia., CC BY-SA 4.0, https://commons.wikimedia.org/w/index.php?curid=113471903

This is because there are fewer ice dams and higher river peak water levels at higher elevations. This is because lower elevations have a greater number of ice dams as well as higher river peak water levels. There is a chance that ice shelves and glaciers are the principal structures responsible for the preservation of freshwater in particular arctic lakes. This is because ice shelves and glaciers are often found near one another. As the temperature continues to rise, there is a possibility that these buildings would collapse, which would lead to catastrophic water drainage if it were to happen. When the Ward Hunt Ice Shelf broke apart in 2002, for example, it caused a lake that had been located on the epishelf for millennia to evaporate and disappear. This was because the lake had been completely isolated from the surrounding environment.

In addition, the ice cover of polar lakes is altered by threshold effects. In both the Antarctic and the Arctic, it is not rare for ice sheets to cover lakes for lengthy periods—years, decades, or even millennia. As a result of the melting of ice, there will be a considerable shift in the limnological gradients. Also, UV exposure increases by orders of magnitude and much surpasses the effects of stratospheric ozone, which leads to a bigger light supply for photosynthesis and more favorable circumstances for the creation of benthic populations. Because of this, there is an increase in the amount of light available for photosynthesis, which creates conditions that are more suitable for the creation of benthic populations. It is possible for the vertical stability of the water column in lakes and ponds found in high latitudes to be compromised as a result of the persistent low temperatures that are found in these bodies of water (Dyck, 2007). During the summer, lakes that are located in colder regions, where the water column is not sufficiently heated to achieve the maximum density of water at 3.98 degrees Celsius, continue to have the ability to mix freely with one another. Higher altitudes are common locations for these types of regions. This barrier is going to be breached, and thermal stratification is going to emerge; as a result of the higher temperature, this is going to create a total shift in the structure of summer.

5.9 BIOGEOCHEMISTRY'S INHERENT LIMITATIONS AS A LIMITING FACTOR

By increasing the quantity of fulvic and humic elements in the soil, planting shrubs and trees may significantly enhance the amount of biomass that is produced by terrestrial plants. This, in turn, causes there to be increased concentrations of particulate and colored dissolved organic material in the waters of lakes. Even relatively minor shifts in vegetation and CDOM concentrations as low as 2 mg L1 have the potential to have a significant impact on the lighting and spectral regime that is found underwater, particularly regarding UV exposure. This is because CDOM concentrations are a measure of how much dissolved organic matter there is in the water. These alterations in the plant life could potentially speed up the biogeochemical transitions that are occurring because they will cause an increase in root biomass, rhizosphere microbial activity, and soil weathering (Domisch et al., 2015). This is because climate change is associated with both a decline in the amount of reflective surface area provided by snow and ice and an increase in the depth of the permafrost active layer. There is also a link between anoxia in the water column and another biogeochemical threshold. If stratification completely depletes the bottom waters of oxygen, large amounts of inorganic phosphorus, iron, and manganese may be liberated from the sediments. There is a possibility that eutrophication will proceed at a faster rate as a result of the increased internal load.

5.10 CONCLUSIONS

Lakes, rivers, and wetland areas that are found at high latitudes are the primary focus of research that is carried out in the field of limnology. This is the case not only because of their immense diversity and significance in polar biomes but also because of their value as appealing models for better comprehending the dynamics of aquatic ecosystems. Plants, animals, and microbes can thrive in these conditions if they have adapted to or are capable of surviving with persistently low temperatures, freeze-thaw cycles, and major seasonal and interannual variations in the supply of energy and nutrition. These conditions can be found in arctic, alpine, and desert environments. These biological communities are today being put to the test by a variety of pressures imposed by humans, such as increased exposure to UV radiation and pollution, which is a direct result of the fact that this issue has arisen in the first place (Dar et al., 2020). Even very little changes in air temperature have the potential to have a sizeable effect on the ecosystems of polar lakes and rivers, which are strongly reliant on ice and the freezing and thawing cycles. These changes might have a significant negative influence on the environment. Because of the roles that they play in the process of global climate change itself, the research, monitoring, and management of these ecosystems are extremely vital.

CHAPTER 6

FOOD WEB DYNAMICS IN INLAND WATER SYSTEMS

CONTENTS

6.1 A DYNAMIC AND INCLUSIVE VIEW OF ECOSYSTEMS

Three aspects of today's pragmatic standpoint of ecosystems are critical for policymakers to understand. To begin with, ecosystems typically have ambiguous and variable limits. They cover a wide range of spatial and temporal scales, nested within each other, from the entire biosphere down throughout its major ecoregions (biomes) to local and potentially fleeting correlations of organisms. Second, human impact on ecosystems is widespread. Humans are seen as an integral portion of social systems instead of as a separate entity. The action of energy, substances, and life forms across boundaries connects ecosystems, and subsidies between terrestrial and aquatic systems are regarded to be especially significant. As a result, anthropogenic impacts in terrestrial systems can have a significant impact on freshwater ecosystems and biota (Cardoso-Silva et al., 2018). Biophysical mechanisms like the transfer of energy, material cycles, competitive pressure, and predation control the complexities of socio-ecological systems, but so do social processes like economics, political systems, culture, and personal choices. Third, environmentalists do not believe that ecosystems are necessarily static and in balance in the absence of human disruption.

Figure 6.1: Yamal Peninsula.

Source: https://www.britannica.com/science/ecosystem

Even in a continuous environment, the comparative magnitude of ecosystem constituents and fluxes varies over time and space due to internal dynamics. Furthermore, attempts to limit this intrinsic variation—or that caused by externally produced disruptions frequently futile, and may harm the ecosystem's ability to adjust to a shifting environment. This is in contrast to the widely held belief that ecosystems have a "natural balance" and gain from being completely shielded from disruption.

6.2 ECOSYSTEMS, ADAPTATION, THRESHOLDS, AND TIPPING POINTS

In environmental policies, ecology, and adaptive biology, the term "adaptation" has various connotations. It refers to human actions designed to reduce adverse consequences in climate policy. Ecosystems are said to be adaptive in ecology because their structure or function can adjust in reaction to a changing ecosystem without requiring deliberate anthropogenic actions. In biological evolution, adaptation refers to a change in the biological properties of a community of individuals as a consequence of biological evolution, a possibility that has become increasingly pertinent to climate change since the Fourth Assessment Report (Cardoso-Silva et al., 2018). The concept of thresholds has emerged as a major ecological as well as political consideration. Three types of limits must be distinguished in order to avoid policy confusion.

The first represents a human preference for the ecosystem to remain within certain boundaries, like above a certain level of forest cover. These, by definition, can be negotiated. The second category reflects basic biological and physiological characteristics, such as the heat flux at which iced soils thaw or species' physiological threshold limits. The 3rd type is characterized by dynamic behavior: the juncture at which the overall effect of all the desirable and undesirable feedback mechanisms governing the system is substantial and favorable that a minor transgression causes a change in the ecosphere state known as a regime shift. In comparison to the previous state, the new state has different dynamic behavior, mean configuration, responsiveness to environmental changes, and streams of ecosystem services. This is known as a "tipping point," and it is significant in the sense of climate change since its onset can be abrupt, difficult to predict accurately, and effectively irreparable. There are now numerous examples of tipping points. Regional-scale ecosystem turning points have not happened recently, but there is conclusive proof for them in the distant past, and there is a major worry that they may occur in the near future.

Early diagnosis and prognostication of ecosystem threshold values, especially tipping points, is a current area of study. There are signs that a rise in ecosystem variability indicates the approach of a threshold. In practice, such signals may be undetectable in the presence of background noise and ambiguity until the threshold is reached. The intricacies of ecosystems are complex, and our current level of knowledge is insufficient to confidently predict all ecosystem outcomes, even if the prospective climate was fully recognized (Wu et al., 2021).

Over the last century, field observations in a variety of boreal, temperate, and tropical ecosystems have revealed biome shifts, or the substitute of one suite of species at a location. Biomes typically move upwards with elevation and to increased latitudes as a result of this effect. Because biome dispersion is known to generally represent climate zones, and the transitions have been witnessed in areas with little human disturbance, such shifts have frequently been credited to anthropogenic climate change. Climate change forecasting of prospective vegetation distribution shows that many biomes could shift significantly, even in areas where environments are largely unaffected by direct human land use. Without a sharp threshold, the magnitude of the shift grows with rising global warming.

Water, as a natural resource, does not constitute "biodiversity." Biodiversity refers to the variety of life that exists. Human impacts on water (whether through pollution or specific use of water to satisfy human necessities) have an impact on inland water ecosystems – and biodiversity reinforces the ecosystem services provided by water to humans. The issues of "water" and "inland water biodiversity" cannot be isolated.

6.3 BIODIVERSITY IN INLAND WATERS

"Inland waters" are aquatic-influenced environments that exist within land borders. This involves those in coastal areas, even if they are close to marine environments. Fresh, saline, or a combination of the two can be used in inland water systems (brackish water). Inland bodies of water include the Caspian Sea (freshwater) and the Dead Sea (hyper-saline), while some would eliminate the Baltic Sea (also primarily freshwater) (Tundisi et al., 2010). Many wetlands in Australia's deep interior, for instance, are saline, whilst most wetlands within close proximity of the ocean are freshwater. It is the location that is important, not the ecological character.

Figure 6.2: Due to the ongoing loss of wetlands, the moor frog has also become rare. In Germany, the species is on the Red List.

Source: CC BY-SA 3.0, https://commons.wikimedia.org/w/index.php?curid=177558

Rather than imposing artificial boundaries, it is preferable to view inland waters through the lens of the larger ecosystem where they are a part. Most coastal aquatic ecosystems are thus best thought of as lower parts of river basins, not least since they are heavily influenced by rivers. Estuaries are areas of transition between rivers and the sea. In practice, "inland waters" considerations appear to concentrate on freshwater – partly because freshwater ecosystems dominate inland waters, but primarily due to the global importance of fresh water (Environmental Agency, 2022). For pertinent coastal areas, the work programs on inland waters and marine and coastlines biodiversity join forces. Lakes, rivers, ponds, waterways, groundwater, springs, cave waters, wetlands, and bogs, marshlands, and swamps are examples of inland waters.

The CBD has embraced the Ramsar Convention's characterization of "wetland." The Ramsar Convention takes a general approach to determining which wetlands fall under its purview. Wetlands are defined as "areas of marsh, fen, peatland, or water, regardless of whether natural or artificial, long lasting or temporary, with water that is still or flowing, fresh, brackish, or salt, which include areas of marine water the depth which at low tide does not surpass six metres."

For the time being, it is adequate to mention that this description encompasses all possible types of inland water bodies or ecosystems, as well as their components, including groundwater. The depth limit only applies to marine areas and not to inland bodies of water (Togaki & Katano, 2020).

Land is included in "inland water ecosystems." All land is an integrated component of an inland water ecosystem from an environmental, hydrological, ecologic, and socioeconomic standpoint because freshwater (typically from rain) starts running off it into rivers, lakes, and wetlands. "Inland water habitat" also refers to land that is directly influenced by aquatic habitat (Environmental Agency, 2022). For example, even if never submerged, the vegetation near bodies of water (in the riverbanks) is significantly influenced by the vicinity to water. Land-water relationships are most visible in seasonally flooded areas, such as river floodplains, which can be clear or submerged based on flood conditions.

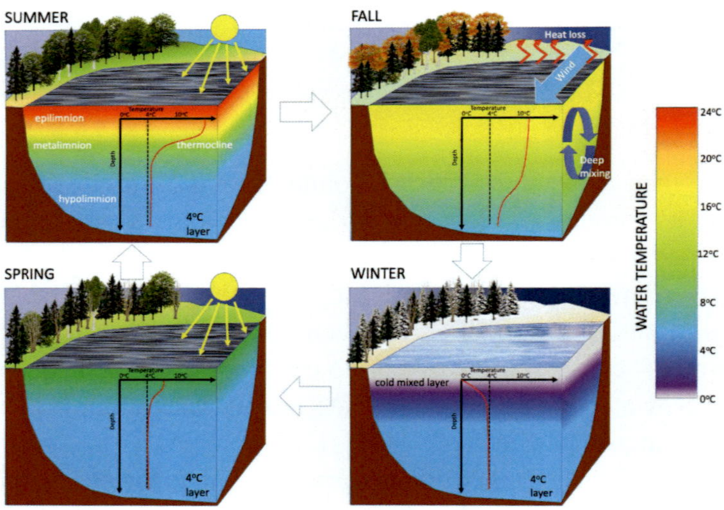

Figure 6.3: Depicts the annual circulation patterns of a dimictic lake. A typical dimictic lake has distinguishable layers that mix completely twice a year. It stratifies in the summer and completely overturns in the autumn and spring. Surface ice inhibits further blending by the wind during the winter. There are minor differences in density and temperature, with cooler water (0 °C [32 °F]) remaining closer to the surface and warmer, denser water (4 °C [39.2 °F]) stretches all the way to the bottom.

Source: By Turbidity.toronto - Own work, CC BY-SA 4.0, https://commons.wikimedia.org/w/index.php?curid=106206428

Ecologically, inland water ecosystems are dynamic. Artificial conceptual boundaries do not apply to them. They are best viewed from a landscape or ecosystem standpoint.

Simply put, it refers to inland water biodiversity. However, because all terrestrial animals and plants rely on fresh water, the lines between aquatic and terrestrial are obscured. At the genetic level, inland water biodiversity includes those life forms that rely on inland water ecosystems for purposes other than drinking (or transpiration in plants). Aside from the apparent aquatic life (e.g., fish), this also includes many "terrestrial" species of animals (e.g., water birds), semi-aquatic animals (e.g., hippopotamus, crocodiles, beavers), and vegetation. For example, most amphibians breed in freshwater (Teffera et al., 2019).

As with all biodiversity, the concept of diversity in inland waters includes species, genetic, and ecosystem diversity. Freshwater fish, for example, cannot easily move between different areas because they are confined to inland waters. Inland waters are thus distinguished by high endemic areas of freshwater species – for example, between various lakes or the upper levels of river sub-catchments, even when they are physically close to each other. This is mirrored in the high rates of genetic diversity. Most notably, ecosystem diversity (such as hydrological and physiological diversity within the terrain) is a critical component of inland water biodiversity (Environmental Agency, 2022). This ecosystem diversity is extremely complex, encompassing both marine and riparian (landscape) influences; preserving it is critical to preserving ecosystem services. Furthermore, anthropogenic influences in the ecosystem actually reduce diversity on purpose (for example, by changing the shape, and thus function, of water courses and/or hydrology).

Inland waters are found everywhere (except in marine areas and, arguably, the polar ice caps, which are made up of freshwater but seldom in liquid form). Inland waters, particularly rivers, are an integral part of all landscapes. On a lesser level, inland water bodies can be found in all terrestrial biomes, as well as grasslands, mountain ranges, forests, islands, agrarian ecosystems, inland coastal zones, and dry and sub-humid lands.

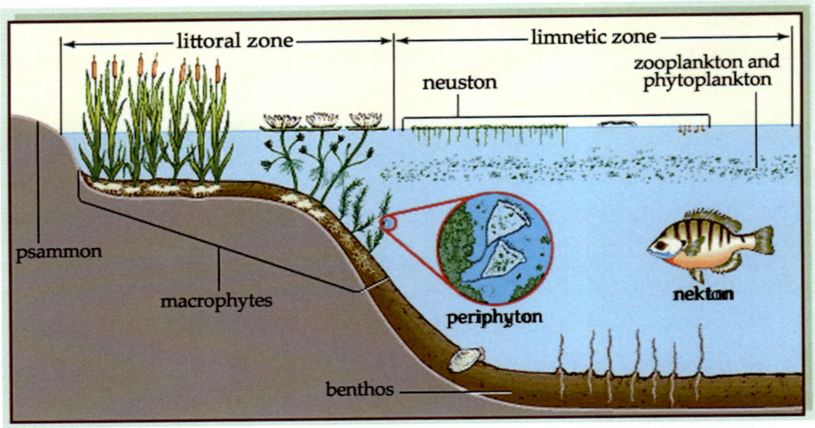

Figure 6.4: Major biological communities of freshwater lakes.

Source: https://www.britannica.com/science/inland-water-ecosystem/Biota-of-inland-waters

6.4 FOOD WEBS

Food webs provide a framework for ecologists to arrange the intricate network of interactions witnessed in nature and across the planet. A medieval Afro-Arab scholar named Al-Jahiz described one of the oldest explanations of a food chain: "All creatures, in summary, cannot live without food, nor can the hunting animal escape being hunted in turn." Lorenzo Camerano created the first graphical representation of a food web in 1880, followed by Pierce and associates in 1912 and Victor Shelford in 1913. Victor Summerhayes, Charles Elton, and Alister Hardy created two herring food webs in 1923 and 1924. In his classic 1927 book "Animal Ecology," Charles Elton spearheaded the notion of food cycles, food chains, and food size; Elton's 'food cycle' was supplemented by 'food web' in a succeeding ecological text. Food webs became a core idea in ecology after Charles Elton used them in his 1927 synthesis (Soranno et al., 2010). Elton classified species into operational groups, which served as the foundation for Raymond Lindeman's quintessential and seminal 1942 paper on trophic dynamics. The concept of a food web has a chronological foothold in Charles Darwin's writings and terminology, such as an "entangled bank," "web of life," "web of complicated interactions," and reference to earthworm biodegradation actions, he talked about "the ongoing motion of the particles of the earth." In 1768, John Bruckner characterized nature as "one continuous web of life."

After Robert Paine's exploratory and descriptive investigation of intertidal shores proposed that food web complex nature was critical to sustaining species diversity and ecological stability, curiosity in food webs grew. This and other discoveries prompted numerous theoretical ecologists, like Sir Robert May and Stuart Pimm, to investigate the arithmetical properties of food webs. A food web is a graphical depiction of what-eats-what in an ecoregion that is formed by the natural connectivity of food chains. The food web is also known as a consumer-resource system. Ecologists classify all living things into one of two groups known as trophic levels: autotrophs and heterotrophs. Autotrophs generate organic matter from inorganic compounds like mineral deposits and gases such as carbon dioxide in order to preserve their bodies, thrive, develop, and reproduce (Environmental Agency, 2022).

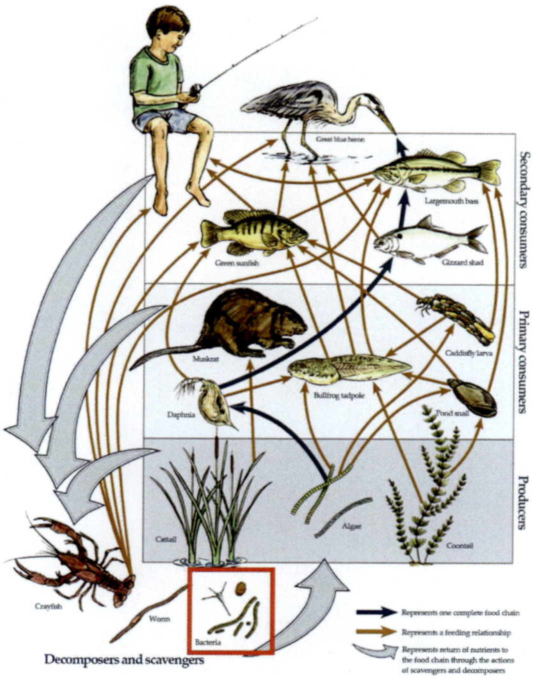

Figure 6.5: Freshwater Food Web-an example.

Source: By Kestin Schulz, Mariya W. Smit, Lydie Herfort and Holly M. Simon. The image (amended) was provided courtesy of the Missouri Department of Conservation. - [1] doi:10.3389/frym.2018.00004, CC BY-SA 4.0, https://commons.wikimedia.org/w/index.php?curid=97329917

These chemical reactions necessitate energy, which is obtained primarily from the Sun and primarily through photosynthesis, with a minor contribution from bioelectrogenesis in wetlands and nutrient electron transfer in hydrothermal ducts and hot springs (Sigala et al., 2017). These trophic levels really aren't binary, but rather form a slope that includes full autotrophs, which get all of their carbon from the environment, mixotrophs (like carnivorous plants), that are autotrophic organisms that get some of their organic matter from places other than the environment, and complete heterotrophs, which must feed to get organic matter.

Food webs are restricted representations of real ecosystems because they must amalgamate many species into trophic organisms, which are functional groups of species in a food web that have the same predators and prey. These simplifications are used by ecologists in quantitative (or mathematical depiction) models of trophic or consumer-resource system dynamics. They can quantify and test for vague trends in the structure of actual food web networks using these models.

Figure 6.6: Animal ecology by Charles S. Elton.

Source: https://openlibrary.org/books/OL13993167M/Animal_ecology

Ecologists have discovered non-random characteristics in the topology of food webs. The published examples used in meta-analysis are of differential quality and contain omissions. Even so, the number of research studies on community webs is increasing, and mathematical diagnosis of food webs employing network theory has revealed patterns that are shared by all (Simões et al., 2011). Scaling laws, for instance, anticipate a relationship between predator-prey linkage topology and tiers of species diversity.

6.5 DIETS AND TROPHIC GROUPS

A descriptive and analytical description of the prey found in a fish's stomach is required to characterize its diet. The most basic assessment method is to note the presence or absence of a specific type of prey in stomachs. The data is then used to compute the percentage of occurrence, which is the ratio of the total count of stomachs containing one type of prey to the total number of stomachs examined (Environmental Agency, 2022).

The numerical method entails counting the number of individuals in each category of prey for the full sample and articulating the total number of prey as a percentage of the total count of prey. Volumetric or gravimetric methods involve determining the corresponding significance of a bracket of prey in terms of volume or weight in comparison to the total volume (or total weight) of any and all prey (Shah et al., 2019).

Subjective methods entail assigning points to every prey based on its presence or supremacy in relation to all digestive systems examined.

Lauzanne proposed a Feeding Index (FI) that takes into account both the proportion of occurrence (percent OC) and the volumetric approach (percent V):

$$FI = \frac{\%\,OC \times \%\,V}{100}$$

Figure 6.7: Feeding Index (FI) formula.

Source: https://books.openedition.org/irdeditions/25217?lang=en

Researchers can use this index, which ranges from 0 to 100, to recognize preferences and evaluate the impact of various prey in the diet. It is beneficial for comparing the dietary patterns of a single species based on size, season, or biotope, and extends itself well to graphical representation (Simmatis et al., 2021).

Following Lauzanne's categorization, a few major trophic categories were identified in Lake Chad, and this characterization can be extended to most riverine tropical African aquatic biota.

Primary consumers that are dominant:

- phytoplanktivorous filter feeders that primarily consume planktonic algae, like *Sarotherodon galilaeus*; (Environmental Agency, 2022)
- grazers and macrophyte consumers like *Brycinus macrolepidotus*;
- browsers consuming the superficial layer of soil particles, which is typically composed of sedimented algae, or periphyton expanding on rocky substrates, such as *Labeo senegalensis*, *Citharinus citharus*, and *Distichodus rostratus*;
- detritivores such as some tilapiines;

Secondary consumers that are dominant:

- zooplanctivores: *Alestes baremoze*, *Synodontis batensoda*, *Synodontis membranaceus*;
- *Tetraodon lineatus*, for example, feeds almost entirely on mollusks
- *Brycinus macrolepidotus* is a surface feeder that primarily feeds on terrestrial invertebrates.

Apex consumers:

- strictly piscivorous fish that feed only on living fish, like *Lates niloticus* and now all Hydrocynus species;
- *Schilbe mystus*, *Bagrus bajad*, and young Hydrocynus are less strictly piscivorous species that eat both fish and shrimp as well as other invertebrates.

In practice, species feed on predators accessible in the ecosystem in which they live, and their diet can vary from one location to another or according to the season. As a result, the preceding classification is only widely applicable (primary, secondary, top). Provided this adaptability to existing funds, West Africa could retain only six broad trophic classifications: limnivores, micro- and macrophytophages, zooplanktophages, invertivores, omnivores, and ichtyophages (Robarts & Zohary, 2018).

Fishes in the East African great lakes have evolved trophic specialties as components of adaptive radiations. Cichlidae appear to use all resources available, and in some cases, exclusively. As a result, we can distinguish several trophic categories:

- detritivores, bottom feeders that consume a mix of planktonic and benthic components;
- phytoplanktivores;
- algal grazers on epilithic (rocks) and epiphytic (plants);
- phytophagous (eats plants);
- molluscivores, which consume shells and gastropods and include oral and pharyngeal crushers;
- zooplanktivorous;
- insectivorous;
- prawn and crab eaters
- piscivorous;
- paedophages consuming other species' eggs and embryos;
- scale eaters who primarily consume scales from other species;
- parasite eaters from outside...

6.6 MORPHOLOGICAL ADAPTATIONS TO FEEDING BEHAVIOUR

If species ended up in equivalent microhabitats in different types of ecosystems, they should have shown similar adaptive mechanisms to identical selection pressure. For diet, scientists enhanced the hypothesis that there might be a relationship between species' general morphology and their diet. Experiments on morphology-diet relationships in African fish are scarce, with many only providing descriptions of the taxa being studied's general diet (Ribas et al., 2019). Even so, a few studies have found that diet and organisms morphology do have a significant relationship. More explicitly, some morphological features appear to be linked to diet type.

These include, for example:

- the mouth and body size of piscivorous fish;
- gut length is related to detritivores' behavior;
- mouth orientation or position describing both benthic and detritivore diets

Findings of the same family or a community consisting of different taxons from different families yielded similar results. They appear to refute the widely held belief that ecomorphological research findings only provide accurate results when restricted to distinctions between families.

6.6.1 Head and Mouth Morphologies

One key advancement, the pharyngeal mechanism, has been proposed as one of the prime causes for the incredible explosive speciation of cichlid fish in the African Great Lakes. The most pronounced morphological specialization involves transformations of the head to collect various types of prey (Environmental Agency, 2022). The extended mandibles of the piscivorous Rhamphochromis and the short, almost rectangular-shaped jaws of the scraper Labeotropheus are particularly poignant examples of evolutionary mechanisms in the jaws.

Many cyprinid species have feeding adaptive responses in the form of their mouth. The tough (or keratinized) lips of the Varicorhinus mouth form are adapted for scratching algae and aufwuchs from underwater rocks or other tough substrates. Labeo's complex ventral mouths, with homodont pharyngeal teeth set close next to each other to create a single grinding area, are specialized for feeding on epibenthic algae (Reed et al., 2008).

6.6.2 Length of the Gut

There is a relationship between diet and gut length in relation to body length. Carnivorous fish typically have a large tummy and a short digestive tract. This is true for Lates and Hydrocynus, as well as many other invertebrate feeders. In sediment browser fish, like Labeo, the stomach is frequently absent or tiny, and the gut is typically more than ten times greater than the body length. In general, omnivorous fish, zooplankton feeders, and invertebrate feeders have intestines that are less than 3 times the size of the body. Moreover, there appears to be no considerable connection between these kinds of diets and gut length.

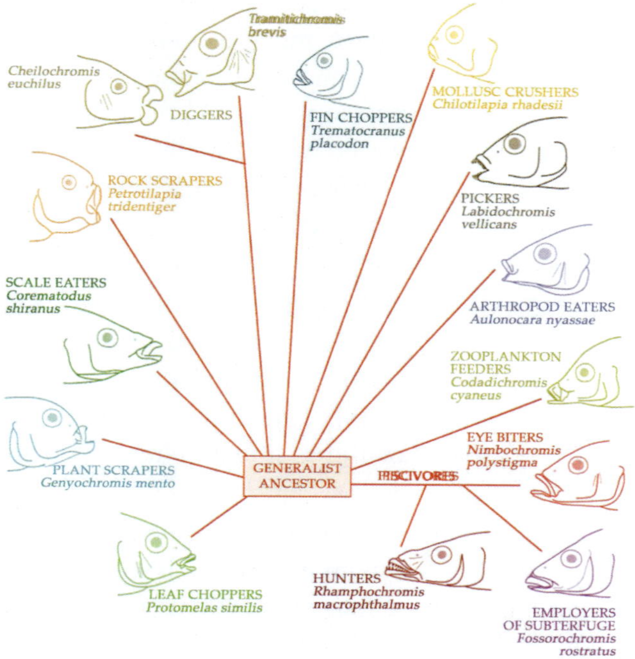

Figure 6.8: Example of adaptive radiation (morphological adaptation of the shape of the mouth) for a few cichlids of Lake Malawi (redrawn from Fryer & Iles, 1972).

Source: https://books.openedition.org/irdeditions/25217?lang=en

6.6.3 Electric Organs

Electric fish have developed unusual adaptations. *Malapterurus electricus* is a slow swimmer, but it sustains almost entirely on fish. Close to a fish, a strong volley of high-intensity electric organ discharges can paralyze it ("Electric organ of *Malapterurus electricus*"). This helps to explain why *M. electricus* is a more efficient fish predator in general (Pinilla & Gabriel, 2016).

Body form and feeding types:

Morphology, which includes mouth size and dimensions as well as body form, constrains foraging behavioral strategies. While searching for correlations between motor function morphology and the feeding niche, Webb identified three basic operational locomotory processes that may be related to foraging:

- In order to exploit food that is widely dispersed in spatial and temporal scales, fish must move at rates of speed that sample the highest volume with the least amount of energy expended. This is true for both macrophages and filter-feeders, which have a streamlined anterior body and a forked caudal fin. *Alestes baremoze* is a typical African fish that screens zooplankton in the shallow marine surroundings and is an outstanding swimmer capable of long-distance migrations (Pinilla, 2010).
- In order to exploit highly mobile food, fish must catch their prey quickly before they could even flee and seek shelter. As a result, they must be ready to speed up quickly. In Africa, a common example is the African pike, that has a versatile body and a very isolated dorsal fin that allows it to accelerate very succinctly but with incredible speed.
- To catch prey that is sometimes concealed in holes or resting on the bottom, fish must swim in very small spaces. The median and matched fins improve maneuverability. Many cichlids, and also mormyrids and notopterids, fit these criteria.

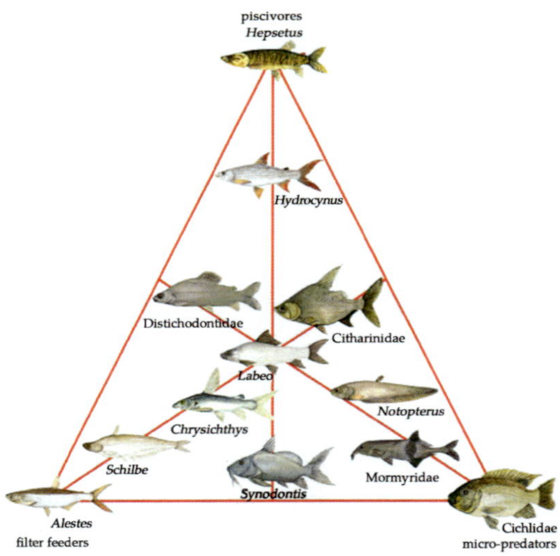

Figure 6.9: The various sorts of fish swimming are represented by a ternary diagram called "functionmorphology." Triangle angles represent three major styles of foraging; species nearer to the center are more generalist.

Source: https://books.openedition.org/irdeditions/25217?lang=en

6.7 PREY SELECTION

The best food source for a predator is ideally one that provides the most energy for the minimal costs of capture in a given context.

Zooplanktivorous fish can feed in two ways (Lazzaro, 1987):

- particle feeders target and consume visually selected individual prey;
- Filter feeders do not discern individual prey but instead collect plankton in their gill rakers. As a result, they engage in passive prey sampling, and efficient filtration is heavily reliant on gill structure and particle size (Pérez-Fuentetaja et al., 2006).

Field workers have used "electivity indices" (E) devised by Ivlev, which are expressed as

$$E = \frac{(r_i - p_i)}{(r_i + p_i)}$$

Figure 6.10: Electivity indices formula.

Source: https://books.openedition.org/irdeditions/25217?lang=en

The fraction of food item ri in the diet, and p I the ratio of environmental item i. This index is useful for determining whether or not a prey species is being ingested in proportion to its ample supply in the ecosystems, but it does not differentiate the mechanisms underlying any selection revealed.

The smallest captured prey in filter feeders like *Synodontis batensoda* in Lake Chad measured 80 m, but the ratio of rotifers and nauplii retained by the gill rakers rises gradually with size, up to 260 m. The ability to avoid predators is crucial in the sampling of the biggest planktonic crustaceans. In this regard, the sizable cladoceran *Moina micrura*, which has limited mobility, is easily captured, whereas the more mobile diaptomids are rarely captured. In the meantime, large *Alestes baremoze* seniors retain only fragments larger than 400 m in their gill rakers, not small nauplii or rotifers (Papa & Mamaril, 2011). All planktonic components are retained above 880 m. In the physical phenomenon where they live in sympatry, these two zooplanktonivorous fish species have different behaviors and do not ingest the same form of prey.

It has also been demonstrated in Lake Victoria that the lowering in the share of shrimp *Caridina nilotica* in the gastric contents of *Lates niloticus* is partly due to the increased spaces between gills as the fish gets bigger. Above 70 cm LT, the gaps are too big to keep the shrimp.

Benthic fishes can also put prey species under selective pressure. Three Lake Chad benthivore species, *Synodontis clarias*, *S. schall*, and *Hyperopisus bebe*, primarily consume small molluscs, as evidenced by a comparison of mollusk size dispersion in the benthos and predator fish gastric contents.

Already in Lake Chad, *Tetraodon lineatus* consumes primarily microcrustaceans and insect larvae, followed by larger lunged mollusks residing in vegetated areas up to 125 mm LS. *T. lineatus* only eats benthic mollusks with thicker shells above 200 mm LS. *Trachinotus teraia*, a lagoon fish, exhibits similar behavior. It sustains itself on Clupeidae larvae and juvenile shrimp up to 125 mm LS, while larger individuals primarily consume *Corbula trigona* bivalves in the Ébrié lagoon.

Figure 6.11: A town on the shores of Lake Chad.

Source: https://en.wikipedia.org/wiki/Lake_Chad#/media/File:LakeChadCameroonTown.jpg

6.8 ONTOGENIC SHIFTS IN FEEDING BEHAVIOUR

Most fish species differ vastly in body size throughout their lives, and their ecology frequently changes dramatically as they grow. Their dietary needs and feeding behavior change as well, which is mainly linked to the capacity of growing animal species to handle specific food types. Numerous research findings on African fish have brought these ontogenetic changes to light. *Tetraodon lineatus* has already been discussed. Instances can also be found in piscivorous species. Juveniles of *Hydrocynus forskalii* are thus strictly zooplanktivorous in the Chari River up to 30 mm in length. They eat both zooplankton and insects and have a body length of 30 to 45 mm. They become piscivorous once they reach 50 mm. The pattern is similar in Lake Chad's Southeast archipelago, but the shift from one diet to another is more common in bigger sizes (Pathak & Mudgal, 2005).

In cichlids, diet shifts with individual development. *Docimodus evelynae*, a cichlid species from Lake Malawi, has a peculiar behavioral and morphological development in which tiny juveniles (less than 50 mm SL) are heterotrophic cleaners, expelling fungi from other cichlids. They aren't cleaners any longer as they reach 50 to 70-80 mm SL and nourish themselves on plankton, insects, and aufwuchs. Adult *D. evelynae* change their diet to feed on the scales, fins, and skin of their fish prey once they reach 80 mm SL. Field data also show that as *D. evelynae* grows, it moves into deeper water. Because many species have depth restricted distributions in Lake Malawi, the cichlids scrubbed by juvenile *D. evelynae* in subsurface water are not like those preyed upon by adults in greater depths.

6.9 RELATIONSHIP BETWEEN PREDATOR SIZE AND PREY SIZE

Many fish have a prey size that decreases the cost of capture by supplying the most energy to the predator. This sort of relationship has been studied extensively in piscivorous fishes, that are the most diverse in this regard. In general, the relative prey size (the size of the prey in connection to the predator) is bigger in small and big predators. As a result, the relationship describes a U-shaped curve.

Figure 6.12: Prey processing by predatory fishes. A) the predator *Cephalopholis urodeta* (SL = 98.4), ingesting a 52.7 mm SL, 27 mm MD *Acanthochromis polyacanthus* in a horizontal, head-first position, and B) a 109 mm SL *Paracirrhites forsteri* that failed to swallow a 45.6 mm SL, 23.1 mm MD, *A. polyacanthus*.

Source: Mihalitsis M, Bellwood DR (2017) A morphological and functional basis for maximum prey size in piscivorous fishes. PLoS ONE 12(9): e0184679. https://doi.org/10.1371/journal.pone.0184679 : https://www.researchgate.net/publication/319605338_A_morphological_and_functional_basis_for_maximum_prey_size_in_piscivorous_fishes/figures?lo=1

Hopson investigated the relationship with both the length of *Lates niloticus* and the size of prey. Lates of all sizes can consume fish measuring up to one-third their own length. The littlest Lates, 30 to 40 mm TL, ate the largest prey in terms of relative size. Relative prey size decreased dramatically from 38% in the littlest Lates to 6% in the 30-36 cm TL group. There was a 6% to 10% increase in the overall size of prey between 300 and 800 mm TL. Prawns and small characins practically disappeared from the diet of Lates over 850 mm TL. As a result, both relative and average sizes have increased dramatically.

The average relative prey size of *Hydrocynus forskalii* in Lake Chad is about 10% for tiny fish (50-75 mm SL) eating mostly invertebrates, but it varies between 20% and 33% for fish from 100 to 400 mm SL. *Ichthyborus besse*, a much less common predator than those previously stated, has also been found to have a correlation between predator size and prey size. The mean relative size of prey varies between 20 and 30% for individual fish just under 100 mm SL and 30% to 40% for *I. besse* above 100 mm SL. The maximum relative prey length reported was 50% that of the predator (Owusu-Boateng et al., 2022).

6.10 VARIABILITY IN DIET OF ADULTS

According to various field studies, many fish may change their diet depending on the season or environment. As the availability of preferred prey decreases, fish appear to expand their diet to involve less preferred prey. Basically when food is limited, individuals are less picky about the quality of their food.

Synodontis batensoda, a detritivorous fish found in the Chari, Nile, and Niger river systems, was solely a filter feeder in Lake Chad prior to the drought. When the food is zooplankton and bobbing organisms, *S. batensoda* swims on its backs to filter the food at the water's surface. If the offered food consists of benthic detritus and silt, the fish swims ventral and filters sediment. During the drying phase of Lake Chad in 1974, when zooplankton was less plentiful, lake populaces of *S. batensoda* displayed both of these feeding behaviors.

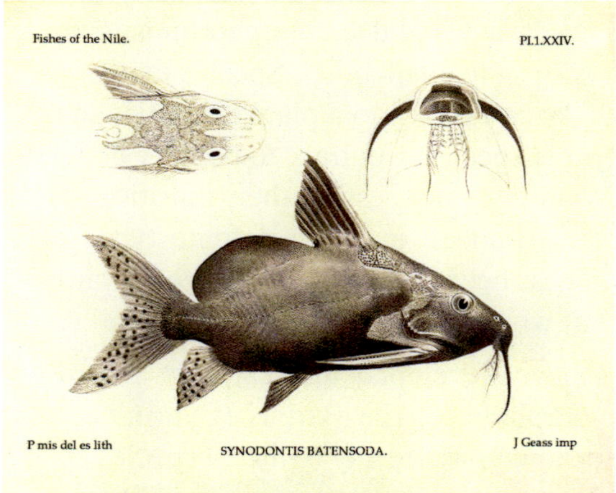

Figure 6.13: *Synodontis batensoda*, the upsidedown catfish.

Source: https://en.wikipedia.org/wiki/File:The_fishes_of_the_Nile_(Pl._LXX-IV)_(6815508218).jpg

In Lake Chad and Lake Turkana, *Alestes baremoze* is a zooplanktivore. In rivers where zooplankton is scarce, such as the Chari and Niger rivers, it becomes largely phytophagous, with a huge element of emerging insects. Aquatic insects and terrestrial invertebrates make up of the diet of *A. baremoze* in rivers in Côte d'Ivoire, whereas this species is a zooplanktivore in the Kossou man-made lake.

More widely, many species modify their diets in response to physical system constraints. It was discovered that for the vast bulk of species, diet commonalities are higher within a sample than that between the same species from different samples obtained from different pools of water of a transitory watercourse in Mali. This suggests that species adapt their diets to the accessible dominant resources in order to survive in somewhat harsh environments. Thus, when faced with low food diversity, which is also associated with lower quality, organisms use the amplest resources, and when prey becomes more abundant, fishes feed on a greater number of prey taxa (Ohtaka et al., 2011). Basically, food diversity, which is dependent on the abundance of ingested prey, appears to be dependent on the availability of prey in the environment.

A substantial body of literature exists on the extent of trophic specialization of endemic Cichlidae in East African lakes. The majority of species appear to be able to consume a much broader variety of food than their morphometric specialization would imply.

In fact, many Malawi Cichlidae are opportunistic and can use one fuel source or another subject to availability. Species that graze on the biological surface layer of rocks can switch to a planktonic diet and even consume the eggs or larvae of other species. Contrary to popular belief, these Cichlidae's trophic specialization does not restrict their nutrition range, but may give them a productive advantage in using definite types of food, especially during food shortages, because they can switch towards a more broad or opportunistic diet at whatever time.

Other groups have developed this capacity for resource sharing by "adapting" their morphology. The Barbus (Cyprinidae) species swarm in Lake Tana is one example, where any of the 14 species analyzed has evolved a unique food strategy to avoid interspecies competition. The morphological research on these 14 species also resulted in the development of an inferential food model (FFM: Food-Fish Model), the results of which are consistent with biological reality seen in the gastric contents of various Barbus.

6.11 FEEDING ACTIVITY PATTERNS

The variation in fish activity over a 24-hour period is a very well occurrence. It is commonly associated with eating habits. Some species are more active during the day, while others are more active at night or during intervals of fast-shifting luminosity (dawn and dusk)

Actions between resting spots and foraging areas, as well as actions to follow, if needed, prey daily activity trends, are examples of major foraging activities. Vertical migrations of Lake Victoria haplochromines, for instance, are thought to pursue the migration of the prey Chaoborus larvae, which spend the day on the muddy bottom and migrate to the water surface at night to feed on zooplankton. Correspondingly, in Lake Kariba, field findings and echo-sounding surveys revealed that the clupeid *Limnothrissa miodon*, which primarily feeds on *Bosmina longirostris*, exhibits a diurnal vertical migration pattern similar to its prey (Naik et al., 2012).

Lunar phases may influence circadian dietary rhythms in species that sense prey by sight. For instance, in *Brycinus leuciscus*, which normally feeds throughout the day, eating activity narrows down during the full moon, despite the fact that feeding is not important at night. In fact, it seems that diurnal feeding lowers due to gregariousness during the migratory process that is guided by the full moon.

6.12 NORTH AMERICA'S GREAT LAKES AND INVASIVE SPECIES

Since European settlement in the region, alien invasive species have posed a threat to the Great Lakes. Since the 1800s, the Great Lakes have become home to more than 140 exotic aquatic species of all kinds, such as plants, fish, algae, and mollusks. With increased human activity in the drainage basin, the rate of exotic species incorporation has increased. Over the last 30 years, more than one-third of the organisms have been introduced, a surge that coincided with the unveiling of the St. Lawrence Seaway. Roughly 10% of the non-indigenous species in the Great Lakes have had considerable economic and environmental impacts. For instance, the sea lamprey has cost recreational and commercial fisheries millions of dollars in losses, as well as millions of dollars in regulation (Mustapha, 2011). Alewife fish riddled beaches each spring, altering food webs and higher water turbidity before exotic salmonids like chinook salmon were crammed as predators and the basis of a new outdoor recreation fishery. Although efforts have not been entirely successful, the Great Lakes Panel on Aquatic Nuisance Species has worked since 1991 to prevent and manage the onset of aquatic alien invasive species in the Great Lakes.

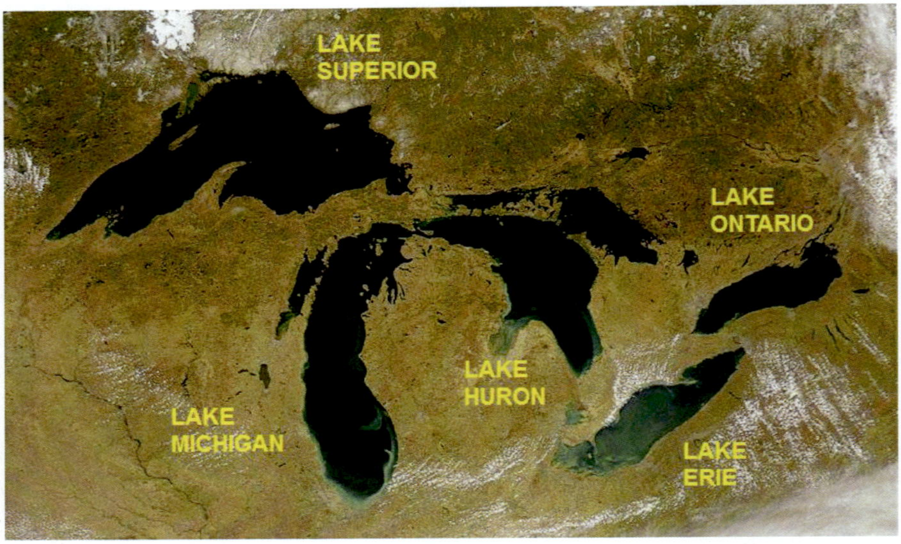

Figure 6.14: Satellite image of the Great Lakes.

Source: https://en.wikipedia.org/wiki/Great_Lakes#/media/File:Great_Lakes_from_space_crop_labeled.jpg

6.13 INVASIVE SPECIES AND EUROPEAN RIVERS

For centuries, canals were built between rivers and other bodies of water in Europe, throughout which species assertively migrated or were assisted by shipment traffic, whether in ballast water or on the hulls of ships. The Volga-Baltic Waterway, which was rebuilt in 1964 and connects the Caspian basin with the Baltic region, is one example of how many aquatic species, including copepods, rotifers, the onychopod *Bythotrephes longimanus*, as well as several fish species, have been relocated to the Volga basin. The Main-Danube Canal, which opened in 1992, has also permitted many Ponto-Caspian invertebrate species to access the Rhine basin and disperse to other basins, primarily in ballast water.

Intentional additions of aquatic species have primarily occurred over the last two centuries. *Gammarus tigrinus*, a North American amphipod, was purposefully introduced in 1957 into the German rivers Werra and

Weser since the local gammarid fauna had vanished caused by excessive chloride pollution. To spur fish production, the mysid *Mysis relicta* has been initiated into many Scandinavian lakes (M Ongaratto et al., 2018). Three North American presented crayfish species have colonized many European waters and introduced the "crayfish plague" (*Aphanomyces astaci*), which has nearly wiped out the native crayfish (*Astacus astacus*).

At least 75 non-European freshwater fish have been initiated into European freshwaters, with nearly 50 establishing self-sustaining populations. When preparations between regions within Europe are included, the quantity of introduced species exceeds 100. The cyprinids and salmonids are the most numerous families, with grass carp (*Ctenopharyngodon idella*), silver carp (*Hypophthalmichthys molitrix*), rainbow trout (*Oncorhynchus mykiss*), and brook char (*Salvelinus fontinalis*) now widely spread in Europe. Only a few additions have resulted in the expansion of economically significant species.

Figure 6.15: The *Dikerogammarus villosus*, also known as the 'killer shrimp' came from water systems in eastern Europe in the 1990s and early 2000s and are terrorizing native species in western European rivers.

Source: By NOAA Great Lakes Environmental Research Laboratory - 1030, CC BY-SA 2.0, https://commons.wikimedia.org/w/index.php?curid=46939080

6.14 PHYTOPLANKTIVORE COMMUNITIES

Simple fish groups feeding on phytoplankton could represent the shortest food chain. One expert , on the other hand, highlighted that the majority of African fish do not connect directly on living plant material, but rather on zooplankton or benthic invertebrates. True primary consumers are found only in a few genera (Labeo, Citharinus, Distichodus, Tilapia, Oreochromis, and Sarotherodon), accounting for less than 7% of all species.

Alcolapia grahami is a cichlid found only in Lake Magadi, where it feeds on benthic flagellated cyanophytes, insect larvae, and copepods. It was presented into Lake Nakuru in the 1950s to control mosquito procreation. It is the only fish species there, and it quickly has become one of the major filter feeders, relying on the cyanobacterium *Spirulina platensis* for 95 percent of its diet (Medeiros et al., 2012).

Lake George has a lot of phytoplankton, mostly blue-green algae. It was primarily ingested by two species, *Oreochromis niloticus* and *Haplochromis nigripinnis*, which accounted for the majority of the fish biomass.

Relatively simple trophic webs: pelagic fish communities of East African Great Lakes:

Because few fish species are adapted to the pelagic zone of inland waters, the trophic structure amongst pelagic fish of East African Great Lakes is fairly simple. In Lake Tanganyika, for instance, the trophic web is primarily arranged by phytoplankton, which is consumed by zooplankton. The latter is the primary food of two clupeid species, *Stolothrissa tanganicae* and *Limnothrissa miodon*, which seem to be narrowly specially designed for pelagic life based on their diet and life origins and are prominent parts of the pelagic food chain, connecting planktonic and piscivorous trophic levels. Limnothrissa juveniles are zooplanktivores, but adults consume juvenile and adult Stolothrissa on a regular basis. The two clupeids are prime prey for predators, which are two principal Lates species. A phytoplankton-shrimp-fish food web is an alternative.

It is worth noting that the pelagic food web of Lake Tanganyika is marine in nature: the primary grazer is a diaptomid copepod, as is the case in many productive aquatic waters, and the main planktivores (Clupeidae) are also of marine sources, as are the primary predators, Lates, which also come from a marine family.

They tried to introduce zooplanktophage *Limnothrissa miodon* (from Lake Tanganyika), the microphytophage *Haplochromis kamiranzovu*, the micropredator *Raiamas moorei* (feeding on invertebrates but also on younger fish), and the piscivore *Haplochromis vittatus* all live in Lake Kivu.

Trophic webs can evolve without necessarily becoming more complex. Lake Victoria is a particularly striking example. After the advent of *Lates niloticus*, the stock of *Haplochromis* spp., which provided as indispensable food for predators, has almost totally vanished in some areas (Matta & Uniyal, 2017). Although certain species continue to survive, they are no longer the primary prey, and the predator, which now frequently exhibits cannibalistic behavior, feeds almost entirely on a single Cyprinidae species.

6.15 MORE COMPLEX FOOD WEBS OF SHALLOW LAKES

Because of the shallowness of the waters, trophic webs in Lake Chad were much more intricate than in the pelagic zones of the great lakes prior to the drought of the early 1970s. In the open waters of Lake Chad, one expert identifies two closely intertwined chains. The very first, planktonic, begins with phytoplankton, while the second, benthic, is partially detritic in origin. Planktivorous Alestida, Mochokidae, and Mormyridae are linked by the planktonic network (*Pollimyrus isidori*). In essence, these fishes are prey to many predator species, which consume only a small proportion of the detritivore and benthic chain species. As a result, zooplanktivorous fishes, as in Lake Tanganyika, are the primary food source for Lates and Hydrocynus. Similar outcomes were obtained. It should be mentioned that terrestrial invertebrates that fall into the open waters of Lake Chad, occasionally tens of kilometers from the coast, play an essential part in the dietary habits of a few species, including *Schilbe niloticus*, which has been observed with digestive systems full of crickets on several instances (Lund & McCullough, 2008). This has also been observed, to a lesser extent, in *Schilbe uranoscopus* and *Bagrus bajad*. Similar findings on *Schilbe niloticus* diet have been produced in Lake Albert. *S. niloticus* contributes significantly to the diet of many other predators in Lake Chad, and serves as the main link in a 3rd trophic chain that links multiple sources to large predators.

What will happen to the multi-species output interplay and population structure understanding of the fish community? Short-term projections, series data forecasts, evidence-based / statistical modeling, and the development of mechanistic / computer simulations are all used. These methodologies are both static and dynamic in order to provide immediate solutions and to connect development with predation and fishing in order to assess yield over time. One major goal is to compare fishing mortality to natural mortality. One goal is to develop predictive models of fish population in order to address "What if...?" scenarios

Here are some examples:

- Models of Generalized Consumption
- Population dynamics linking predator and prey populations.
- Models of predator-prey interactions.

The present system of this continental shelves region's fish community has evolved in response to multiple disturbances (e.g., harvesting and habitat breakdown) over time. Why has the fishing community reacted in this manner? With an exploratory investigation of several strategies validated with data collected, There have been general trends in fish trophic interactions that relate to basic ecological principles.

CHAPTER 7

PLANTS THAT LIVE ON, IN, OR NEAR INLAND AND MARINE WATERS

CONTENTS

7.1 INTRODUCTION

Figure 7.1: Plants near rivers.

Source:https://lmg-labmanager.s3.amazonaws.com/assets/articleNo/23368/aImg/43314/july23-2020-nioz-river-plants.jpg

A plant is considered to be native to a wetland environment if it is able to thrive in or on water, as well as in soils that have been submerged or soaked for a lengthy period. Wetland plants can flourish thanks to the presence of anaerobic conditions in the root zone. The flora of a wetland habitat may be identified quite easily since it is mostly made up of plants that are native to wetland environments. The classifications of hydrophytes, macrophytes, and aquatic plants are all applicable to these plants. The majority of the plants that may be found in wetlands are angiosperms, sometimes known as flowering plants. It has been estimated that between three and five percent of the two million angiosperm species are suitable for growing in wetland environments (Leipe et al., 2014). The capacity of some ferns, such as the *Ceratopteris pteridoides*, sometimes known as the Floating Water Fern, to float allows them to thrive in wetland environments. *Taxodium difficile*, *Larix laricina*, and the south Florida slash pine (*Taxodium floridum*) are examples of trees that can thrive in marshy environments (*Pinus elliotti*). Peatlands are the habitat of bryophytes belonging to the genus Sphagnum, which includes several different types of mosses. The majority of our time will be devoted to a conversation on angiosperms that are found growing in marshes.

There is a large variety of adaptations, ecological tolerances, and life cycles among the many species of wetland plants; nonetheless, the ability to live in water is the most important and defining characteristic. Even though wetland plant species and their terrestrial counterparts share many of their adaptations, there are a number of these properties that are either exclusive to one species or have evolved to a high level of specialization when present in numerous plant species.

Wetland plants are responsible for a variety of important functions in wetland ecosystems, including the following:

- Because of their position at the bottom of the food chain in wetland environments, they are a very important source of energy. Nevertheless, many herbaceous wetland habitats attain production levels that are comparable to those observed in tropical rain forests. This is the case even though primary productivity in wetland ecosystems is very different from that of forest ecosystems.
- Numerous taxonomic taxa, such as bacteria, epiphyton (algae that grow on the surface of plants), macroinvertebrates, fish, and birds, are reliant on wetland habitats for their survival. The composition of the plant community that inhabits the wetland has a role in determining the degree to which the wetland's overall variability can vary (Larsen & Alp, 2015).
- The plants that are native to wetland environments have a significant impact on the elemental make-up of the water. This is because these plants act as nutrient sinks and pumps for the surrounding environment. They transport the compounds that are found in the sediment to the upper layers of the water. They contribute to an improvement in the quality of the water by absorbing nutrients, metals, and a variety of other impurities, which they do so.

Figure 7.2: River plants.

Source:https://www.canr.msu.edu/contentAsset/image/63fa4f47-8cd0-48a7-9f13-afeb3ad170b7/fileAsset/filter/Resize,Jpeg/resize_w/750/jpeg_q/80

Because they assist to construct shorelines, controlling currents, and reducing the amount of damage that is caused by floods, the plants that grow in wetland areas have a significant influence on the hydrology and sedimentation processes that occur in wetland areas. The amount of peat that builds up, the volume of water that is cast into shadow (which affects the temperature of the water) and the amount of water that is lost via transpiration are all examples of how vegetation may alter the hydrologic conditions. For instance, wetland plants are one of the tools that wetland managers and researchers use to keep wetland ecosystems in good condition and to manage them effectively (Kumar, 2022). The United States of America, along with many other nations located all over the world, makes extensive use of the plants that are native to wetland areas as a means of assisting in the identification or description of the jurisdictional limits of wetlands. It is possible to use the composition of the plant community that inhabits a wetland as a biological indicator of the 'health' of the wetland.

This may be done in several different ways. It is possible to put this sort of data to use for a variety of purposes, such as keeping track of the state of wetland areas throughout the course of time and developing restoration goals for wetland areas. The lifespan of individual inland water bodies can range from days to tens of millions of years, which is a significant amount longer than the lifespan of inland water ecosystems, which have most likely been present for as long as the continental crust has been exposed to the atmosphere. This makes the lifespan of inland water ecosystems significantly longer than the lifespan of individual inland water bodies. When compared to marine or terrestrial environments, the physicochemical diversity of inland water settings is significantly greater. This applies to photolithotrophic organisms. This is because the spectrum distribution of down welling radiation is bigger in inland water environments. Additionally, the concentration range of specific solute species is greater in these settings. Both of these factors are responsible for this phenomenon. Over the past 500–1000 million years, there has been a significant rise in the rate of weathering and the productivity of terrestrial biota. As a result, there has been an increase in the number of chemical inputs to inland streams.

These inputs are the result of biological activities that have taken place. The diversity of photosynthetic species found in inland waterways is significantly higher than that found on land, but it is significantly less than that found in oceans at both the Phylum (Division) and the Kingdom levels. However, the diversity of photosynthetic species found in inland waterways is significantly higher than that found on land. Every planktonic species that play a role in photosynthesis in freshwater systems originated in marine environments. In contrast to the pleustophytic, rhizophytic, and haptophytic embryophytes that are generated by benthic algae, pleustophytic, rhizophytic, and haptophytic embryophytes are produced by terrestrial plants (bryophytes, tracheophytes) (bryophytes, tracheophytes). Changes in biomass, net gas exchange, tracer consumption (primarily inorganic C), and chlorophyll fluorescence measurements have all been used to estimate primary production in inland waterways, with chlorophyll fluorescence measurements serving as the primary focus.

Other methods, such as net gas exchange and tracer consumption, have also been utilized. Each of these strategies has several benefits as well as drawbacks. The tactics point to a distinct set of results as a consequence of adjusting the light supply (mixing depth, light attenuation), in addition to the accessibility of nutrients. This is because the light supply is being changed. Eutrophication, which is caused by an excessive supply of nitrogen and phosphorus, results in an increase in primary production at the expense of biodiversity. It also has the potential to result in a reduction in the amount of carbon dioxide that is exhaled by inland rivers. The use of fluorescence techniques can make some important contributions to the investigation of primary productivity in inland waters (Knight et al., 2015). Some of these contributions include the continuous in-situ measurement of algal fluorescence as a proxy for biomass, remote sensing of chlorophyll fluorescence, and further development of fluorescence methods for estimating primary productivity. These are some of the most important contributions.

The temperature of the water in a stream can harm the ability of aquatic biota to thrive and survive. However, the presence of forest cover on stream catchments helps to prevent rivers from warming up. This is beneficial to aquatic biota. There was a correlation between the level of cloudiness in the sky and the temperature of the air, which in turn affected the temperature of the water. An increase in the amount of solar radiation that enters the stream catchment zone as a result of the removal of plants, particularly those with enormous leaves, is possible. This can also lead to an increase in wind speed and air exposure, both of which can contribute to an increase in air, soil, and stream temperatures during the summer while simultaneously contributing to a drop in relative humidity.

Both of these effects can be caused by the same thing: Temperature increases in streams and, as a result, the removal of plants is influenced not only by changes in separation but also by the hydrology of streams and the structure of their courses. Separation plays a smaller role in this process. Recent reports from Iraq suggest that environmental conditions have been deteriorating and that there have been shifts in the surrounding environment (Knott et al., 2002). The environment of the city of Basrah, which is located in southern Iraq, is a good illustration of such severe environmental changes. Basrah is located in the country of Iraq. Some of the changes that have been seen and recorded include desertification, secondary salinization, urbanization, vegetation deterioration, and the elimination of wetland zones. It was suggested that the streams and the catchments that surround them be restored for any plan of development led by the government to be carried out. This is because the streams are what feed the catchments.

Biota of Inland Waters

A surprisingly diverse collection of plants, animals, and microorganisms reside in inland waters, with essentially all significant groups of life forms traced back to one kind of aquatic environment or another. However, no significant group really developed in inland waters; all came either from the ocean or lad, whence the biological infiltration of inland waters ultimately occurred. The extensive time duration since this invasion happened, notwithstanding, has permitted different significant taxa of inland waters, like various kinds of crustaceans, to develop. The main significant groups of aquatic animals prominently missing from inland waters are the phyla Ctenophora, Hemichordata and Echinodermata.

Among the animals include the phyla Cnidaria, Porifera (sponges) and Bryozoa (moss fauna) and among the plants are Rhodophyta (red algae) and Phaeophyta (brown algae). The explanation on why these groups didn't infiltrate as much as other groups is unclear, yet apparently, they were less ready to adapt to lower salinities and diminished environmental stability. Significant groups of inland aquatic biota that are derived from terrestrial ancestors are insects and macrophytes apart from large algae.

Despite their origins, the invading biota needed to foster variations to the unique physicochemical conditions of inland waters. For those leaving a marine ecosystem, the original transformation was physiologic and that would allow endurance in a significantly less saline, diluted outer medium.

7.2 SEA PLANTS

Figure 7.3: Sea plants.

https://media.istockphoto.com/photos/seagrass-underwater-sunlight-mediterranean-sea-picture-id1168644495?b=1&k=20&m=1168644495&s=170667a&w=0&h=AMAwzSsIqdM6A-QSIuLXgaBn4da5OiBwfxMQ2p_JgQU

It is possible to locate aquatic plants that can flourish and develop regardless of the environment they are in (or areas). Some species prefer bottoms that are sandy or muddy, while others need rocks, piers, reefs, or other kinds of marine flora as a means of attachment to stay in place. Some individuals like sites with calm waters, while others choose sites with more turbulent waves. There is a certain temperature range required for the growth of seaweed. Because of the significant temperature differential in the water, the marine plants that can be seen flourishing in the Pacific Ocean are distinct from those that can be found growing in Japan and New Zealand. Only one species, Cladosiphon, can only be found in the cooler waters of Tonga and New Caledonia; the remainder of the sea plants described in this book may be found in a wide variety of locations across the Pacific Islands.

The depth of the water, the amount of salt in the water, and the amount of light that is present all have a role in the proliferation of marine plants. The vast majority of species can only live in saltwater with a salinity level that is higher than freshwater around the mouths of rivers, but there are a few exceptions. When the tide is out in the open air, some individuals can remain calm and collected despite being subjected to conditions like driving rain or intense sunlight. Those that are delicate, on the other hand, ought to be submerged in salt water at all times. Others may be found hidden beneath the surface, in the depths of the water, behind rocks or other types of plants, or even in the water itself (Kazanci et al., 2004). Brown or green sea plants, on the other hand, are more usually found in shallow water, whilst red sea plants are more likely to be found in deeper seas than brown or green sea plants.

Both aquatic and terrestrial plants follow a variety of growth patterns during the different seasons. Not all of them emerge just during certain times of the year, thus this does not mean that they are the only ones. In fishing communities, the information about where and when to acquire sea vegetables is frequently passed down from generation to generation. It is common practice to keep the most desirable locations a secret.

7.3 ALGAE

The word "algae" is used to refer to a more generic category that encompasses both "seaweeds" and "sea plants," even though these creatures are referred to as "seaweeds" and "sea plants" in a more scientific meaning. The shallow waters that surround the world's oceans are home to a diverse variety of marine plants, which are commonly referred to as sea plants. Sea plants are found in every ocean on the planet. There are many different kinds of organisms that make their homes in the seaweed and phytoplankton that are found in the ocean (Jeong et al., 2008). Not only do these creatures contribute to the aesthetic appeal of the undersea environment, but they also play a vital role in the food chain and are directly significant to people as sources of food and raw materials for industrial use. The process of converting carbon dioxide and water into food for the organisms that live within marine plants is termed photosynthesis.

This process, which is dependent on the rays of the sun, is carried out by marine plants. In terms of nutritional requirements, algae are capable of satisfying their demands (primary producers) (primary producers). Oxygen is produced as a byproduct of this process, which ultimately results in the generation of oxygen itself. In the context of marine ecosystems, algae are categorized as "plants" and serve an important function. The vast majority of terrestrial plant species are more complicated to comprehend than their marine counterparts due to the presence of roots and branches in their structures. Sea plants, on the other hand, lack both of these components. Because they acquire the nutrients they require directly from the water that is all around them, sea plants do not develop roots or any other sorts of complex conducting tissue. Some of the world's largest marine plants, such as kelps, include holdfasts, stipes that look like stems, and fronds that look like leaves. One of the criteria that is taken into consideration while determining the genus that an alga belongs to is its color. There is green sea vegetation, brown marine vegetation, and red marine vegetation. Some sea vegetation is red, and some are brown.

The presence of photosynthetic and accessory pigments, which each reflect a specific color of light and give green, brown, and red sea plants their distinctive appearances, is the criterion that is used to place these organisms into one of three categories. The green, brown, and red sea plants each reflect a different color of light. There is a noticeable difference in the hue of light reflected by the green, brown, and red sea plants. Photosynthesis can't take place in the absence of the pigment chlorophyll-a, which is present in all marine plants to some extent. It is generally accepted that algae were the very first photosynthetic cellular "plants" ever to exist, and as a consequence, it is believed that all subsequent plant groupings descended from algae. Algae were termed plants (Islam & Chowdhury, 2013). They are a varied group of organisms that can be unicellular or multicellular in size, exhibit minimal thallus differentiation, and lack the key conducting tissues that are present in mature plants. The size of these organisms can range from unicellular to multicellular. They can adapt to a wide range of conditions, including those found in freshwater, marine, and even terrestrial settings.

The structure of macroalgal thalli may be what determines whether or not they are parenchymatous or filamentous. These two different classes each have their own set of benefits and drawbacks. The name "kelp" is used to describe algae that have holdfasts, stems, and fronds that may grow as tall as trees, whereas the term "seaweed" is used to refer to bigger algae that have a thallus (for example, Macrocystis can reach a height of 65 meters). There are a few distinct kinds of algae, including seaweed and kelp. Seaweed is a kind of algae that is much bigger and has a thallus. Kelp is a kind of algae that is much smaller and does not have an (e.g., Macrocystis reaches 65 m height). Massive populations of kelp in tropical places generate "forests" that are kept in deep waters by nutrient-rich upwellings of cold marine currents. These "forests" are referred to as "sea forests." The canopies of these forests, which are prevalent in tropical locations, protect the algae that are located below them (like Laminaria, for example), and these forests can be discovered (e.g., Laminaria). Green algae (Chlorophyceae), brown algae (Ochrophyta, Phaeophyceae), and red algae (Rhodophyta) are the three types of seaweed that can be differentiated from one another based on the colour of their thallus as well as the amount of storage capacity that it provides (Chlorophyta). Red algae are classified as Rhodophyta (Ishida et al., 2020). Brown algae are classified as Ochrophyta. Green algae are classified as (Ochrophyta, and Phaeophyceae).

Red algae (Rhodophyta), brown algae (Ochrophyta, Pha (Chlorophyta). In addition to asexual and sexual reproduction, their reproductive activities involve the fragmentation of vegetative thallus tissue, the creation of motile zoospores or gametes, and, in certain rare cases, the alternating production of haploid and diploid progeny. Seagrasses are a polyphyletic group of monocotyledonous flowering plants that grow submerged and have their roots in marine and estuarine soils. Seagrasses are one of the most abundant types of plant life in the ocean. There are a few other names for seagrasses, including eelgrass and emerald grass. There is a relationship between algae and seagrasses, and both of these organisms are classified in the same kingdom. Because there are only fifty to sixty species, thirteen genera, and six families (Hydrocharitaceae, Ruppiaceae, Posidoniaceae, Zosteraceae, Cymodoceaceae, and Zannichelliaceae) in the entire world, these plants are not recognized to be real grasses.

It is necessary for seagrasses to exist in an estuarine or marine environment, to have adapted to high and possibly fluctuating salinities, to be able to grow completely submerged, to have specialized leaves that do not have stomata and a reduced cuticle, to have an underground stem and roots that resist anoxia and anchor the plants, and to have a submerged pollination mechanism. All of these requirements must be met for seagrasses to exist. Seagrasses can't exist unless each of these conditions is satisfied. Additionally, for seagrasses to properly reproduce, they require a method of pollination that occurs while the plant is submerged. For their survival, seagrasses need the ability to propagate (Hodgson et al., 2001). It is important to note that seagrasses do not exhibit the vast majority of the traits that are described in the following paragraphs.

Figure 7.4: Algae.

Source:https://www.gannett-cdn.com/presto/2021/07/14/USAT/b7768369-7a61-4e95-80e4-119aeb6278e4-GettyImages-1164651319.jpg

It is generally accepted that algae and seagrasses make up the foundation of the food web. They are the principal contributors to the synthesis of organic compounds in the water, which positions them at the very bottom of the food chain. Because of this, they are at the bottom of the food chain. In addition to this, they may construct habitats, which would lead to changes in the ecology of the region and a reduction in the number of diverse marine animals. This would be a negative impact. Seagrasses stop the buildup of silt, which is a necessary ingredient for the formation of macroalgae but may be prevented by their presence. Coral reefs and the rhodolith deposits that are created by calcified macroalgae are particularly significant because they provide homes for a diverse range of plant and animal species (Helm, 2017). This makes coral reefs and rhodolith deposits extremely essential. Because of the positive impacts that these species have on the habitats in which they live, they are sometimes referred to as "ecosystem engineers." Epiphytic creatures can exist on the surface of their host, as endophytes within their thallus and tissues, as epilithic animals near to their thallus, or as episamic organisms farther away from their host. Epiphytic organisms can also exist as endophytes within their thallus and tissues. It's possible for organisms that live on plants' surfaces to also live as endophytes in the plant's thallus and tissues. There is no denying the significant role that macroalgae play in marine ecosystems; yet, the fact that these organisms have begun to colonize areas of the ocean that are not home to coral reefs raises some cause for concern. In addition to microalgae, the phrase "seaweed" can apply to macroalgae.

7.4 MACRO ALGAE AND SEAGRASSES

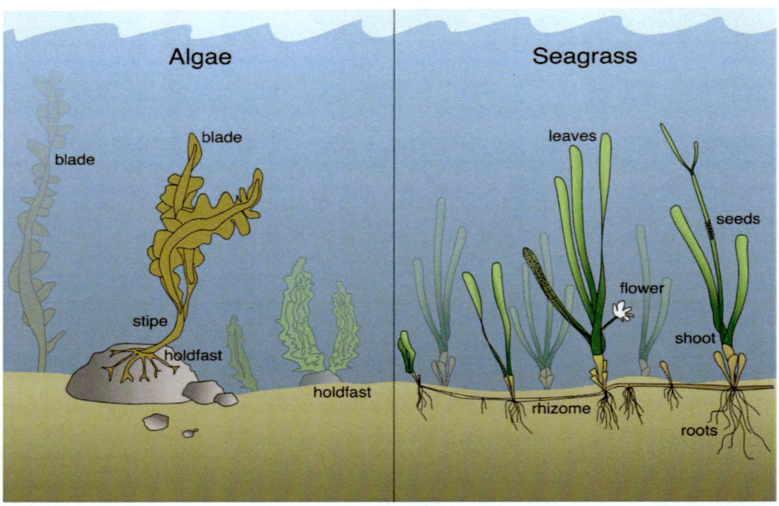

Figure 7.5: Macro algae and seagrasses.

Source: https://pbs.twimg.com/media/DpKzggJWwAAtz_B.jpg

There are many different kinds of tropical habitats, such as coral reefs, beach rocks, rocky shorelines, sandy beaches, estuaries, mangroves, and seagrass beds that provide a variety of substratum and safe havens for algal spores to land on and grow in. Even though their environments are more diverse, the ecological diversity of tropical macro algae is not on par with that of warm-temperate zone warm-temperate zone warm-temperate zone warm-temperate zone warm-temperate zone warm-temperate zone warm-temper the lack of biodiversity in tropical regions may be caused by competition for space with hard corals. This may be because hard corals limit the number of marine algae that can grow in large areas where biogenic reefs have developed (Hassan et al., 2018). The presence of herbivorous fish and sea urchins, in addition to the low quantities of nitrogen that are present in the saltwater, are to blame for the observed decline in the variety of algae that can be found in tropical seas.

Likely, the reason why only a relatively small number of species of algae have been described is connected to the fact that surveys of the marine flora in tropical regions are carried out on a much less frequent basis than they are in other parts of the world. The Eastern Atlantic, the Western Atlantic, the Indo-West Pacific, and the Eastern Pacific are the four key regions in which tropical marine algae can be found. The diversity of algae in the Caribbean is forecasted to approach 600 species, while the diversity of algae on the northeastern tropical coast of Brazil is forecasted to be at least 400 species. Both of these regions can be found in the western part of the Atlantic region. At least 300 distinct species have been classified as living along the coast of Africa's western coast. The tropical waters of the Indo-West Pacific make up the largest tropical marine region on the planet and are home to the greatest variety of tropical marine life of any region on Earth. For example, researchers in the Philippines have compiled a list of close to 900 unique species of marine algae (250 species of green algae, 154 species of brown algae, and 506 species of red algae) (250 of green, 154 of brown and 506 of red algae). Because there are probably fewer than 300 infrageneric taxa in the Eastern Pacific, this region has the least amount of species diversity of any area in the world (Gothwal, 2021).

Although it is possible to find seagrass in most of the shallow coastal waters around the world, the majority of seagrass species are only found in tropical or temperate zones. Seagrass can be found in most of the shallow coastal waters around the world. The Greater Caribbean, the Tropical Indo-Pacific, and the South-East Atlantic/West Africa zones are the only three of the nine to ten entirely tropical seagrass zones. Seagrass zones can be found in a wide variety of depths and temperatures. The remaining seagrass zones are sectioned off according to the types of plants that grow in those areas. The limited number of seagrass species that are truly adapted to temperate conditions and the overwhelming majority of seagrass species that are adapted to tropical climates are two factors that accurately describe the distribution patterns of seagrass. Southern Australia and Japan, two of the five places with growing seagrass diversity, both have temperatures that are more temperate than the temperatures of the other three tropical regions (Southeast Asia, eastern India, and eastern Africa). The most diverse collections of seagrasses can be found in India, the Philippines, and Papua New Guinea, which each have a total of twelve species of seagrass.

India has fourteen, the Philippines has thirteen, and Papua New Guinea has twelve species (12 species). These patterns are remarkably similar to those that have been identified in corals, mangroves, and macroalgae; however, the number of species that make up the algal flora is significantly larger than any of the other flora types. The Caribbean is home to a diverse community of seagrasses, the most common of which are Halodule, Halophila, Ruppia, Syringodium, and Thalassia. However, there are also a few other species of seagrass. The Brazilian flora is still considered to be a part of this seagrass zone, even though it is physically separated from it by the estuary of the Amazon River. The presence of the species Halodule, Halophila, and Ruppia is responsible for this distribution (García-Girón et al., 2020). The Cymodocea, Enhalus, Halodule, Halophila, Syringodium, Thalassia, and Thalassidendron genus families are among the most well-known genera that can be discovered in the Tropical Indo-Pacific region. *Halodule wrightii* is the only species of Halodule that is known to reside in the area encompassing the South-East Atlantic and West Africa.

The seagrass ecosystems of the Mediterranean and South Africa are neither temperate nor tropical, but they do have both tropical and temperate species living within them. In the marine algal flora, the number of red algae (Rhodophyta), brown algae (Ochrophyta, Phaeophyceae), and green algae (Ochrophyta, Phaeophyceae) are classified according to the pigment concentration in their thallus (Chlorophyta). Marine algae with red and green pigmentation are most common in tropical environments, whereas brown algae are found in greater diversity in cold temperate regions. On a latitudinal gradient in the northern hemisphere, the ratio of red to brown algae increases from high to low latitudes from 0.6 to 3.9 in the Western Atlantic Ocean and from 2.4 to 4.6 in the Eastern Pacific Ocean. This finding indicates that there is a preponderance of red algae species in tropical regions.

The proportion of red algae to brown algae decreases from high latitudes to low latitudes in the Southern Hemisphere, going from 0.6 to 3.9 as one moves from high to low latitudes. Even though there is a large diversity of red algae over the biogenic reefs and rocky shorelines of the tropical Atlantic, seasonal coastal upwelling brings cold water to the surface, which makes it possible for a variety of brown algae species that are more suited to temperate climates to coexist. This is because brown algae prefer cooler temperatures. This ratio falls to 3.2 in areas where upwelling occurs, such as the region close to Cabo Frio, but it remains at 4.7 throughout the entirety of Brazil's tropical coast in general, where it averages 4.7. Both variations in temperature at different latitudes and geographic isolation can play a role in helping to explain the spread of a wide variety of different types of marine algae (Gagan et al., 2018). A geographical barrier between the Atlantic and Pacific seas was produced as a result of the land bridge of Panama, which was produced as a result of continental shelf movements that occurred throughout the Tertiary period of geological time. These movements took place.

As a result, the algae in these two bodies of water were dissimilar from one another. Algae can attach themselves to ship hulls and be transported through the Panama Canal; however, the presence of herbivorous fish makes it difficult for algae to thrive in the area. The Suez Canal, which is a passageway that links the tropical Eastern Atlantic with the tropical Indo-West, is the second challenge that must be overcome. Because they migrate there, certain types of algae are only found in the tropical regions of the Indo-Pacific and the Caribbean. These regions have a tropical climate. This migration could take place through the Suez Canal in the Mediterranean or around the Cape of Good Hope in South Africa. As a result of the movement of ocean currents such as the Equatorial Countercurrent, numerous new species of algae have been discovered all across the Atlantic Ocean, from the coasts of the United States to those of West Africa. These discoveries were made in the Atlantic Ocean.

A number of the same genera and species of algae can be found throughout the tropical seas of the entire planet. A pantropical distribution is another name for this kind of distribution. Numerous examples can be found in the genera Dictyota, Dictyopteris, Padina, Sargassum, and Turbinaria. These genera are all members of the orders Caulerpales, Cladophorales, and Dasycladales. Brown Dictyotales and Fucales are also included. Each of these orders is considered to be a type of algae. Rare and endangered species are frequently restricted to being found in only a few specific locations across the globe. This may occur as a result of factors such as geographical isolation or unusual oceanographic conditions. Because of the presence of upwelling waters in tropical regions, the number of species that can flourish in warm oceans is restricted (Eaton & Kardos, 2015). The possibility that the dispersal of seagrass could lead to the existence of "twin species" in different regions is an interesting and exciting aspect of the process. These are species that are very similar to one another and can be found in a few different types of flora.

The *Thalassia testudinum-Halodule wrightii-Syringodium filiforme* group from the Caribbean is paired with the *Thalassia hemprichii-Halodule pinifolia-Syringodium isoetifolium* group from the Indo-Pacific. It is quite likely that the migration of continents was the cause of the creation of these paired assemblages. Even though it co-occurs with outer Halophila species that have more restricted geographic ranges, the distribution of *Halophila decipiens*, a tiny seagrass species that live in deeper water, is the most widespread in tropical seagrass zones. This is because *Halophila decipiens* prefer to live in deeper water.

7.5 OCEAN PLANTS

Common Types of Ocean Plants

There are two categories of marine plants that may be discovered in the ocean. These categories include floating plants and submerged plants. Plants that grow in the water yet rise to the surface to float are referred to as floating plants (Degefu et al., 2014). On the other hand, rooted plants can only flourish in the presence of sunlight, which is why they like the more shallow coastal waters around the shore. The following is a list of the four primary classifications of marine plants, both those that float and those that are rooted;

7.5.1 Phytoplankton

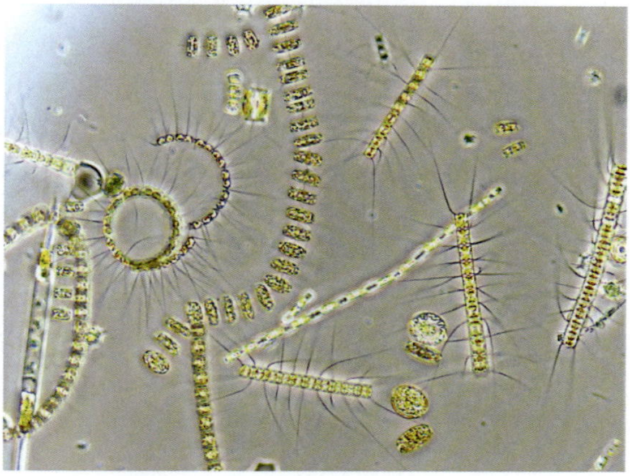

Figure 7.6: Phytoplankton.

Source: https://upload.wikimedia.org/wikipedia/commons/thumb/7/7b/Mixed_phytoplankton_community_2.png/1200px-Mixed_phytoplankton_community_2.png

There is a possibility of having autotrophic, or self-feeding, phytoplankton in both saltwater and freshwater environments. The phytoplankton community is by far the most abundant type of plant life in the oceans, which are home to a diverse range of plant species. It is a term that refers to a vast variety of plants that may be found thriving in the water, such as algae and diatoms. It is the interaction of sunlight with water that allows for the growth and reproduction of microscopic organisms known as phytoplankton. However, even though phytoplankton species only prevail in marine environments, they are one of the most extensively dispersed groups of creatures on the planet. They leave behind a legacy for subsequent generations to continue the task that they began once they are gone. They had such a huge population that they believed it was important to take this action to assure that they would continue to exist (Dyck, 2007). As a method for producing energy, photosynthesis is utilized by phytoplankton, just as it is by trees and other land plants. It should not come as a surprise to find phytoplankton in bodies of water that receive more light than those that do not, given that these organisms require sunlight to live (euphotic zone). For instance, phytoplankton is dispersed throughout a far larger area, and their rate of succession is significantly faster than that of terrestrial plants like trees (days versus decades). As a direct consequence of this, phytoplankton is very sensitive to variations in the climate of the entire planet.

Because phytoplankton plays such an important role in the global carbon cycle, marine and freshwater food webs alike rely heavily on them as a primary source of nourishment. Even though algae only make up a fraction of one percent of all plant life on Earth, they are responsible for producing at least half of the oxygen that we breathe. The process of photosynthesis carried out by algae accounts for approximately half of the world's total oxygen production. Because of their capacity for photosynthesis, coccolithophores, which are characterized by armor plating, constitute the vast majority of phytoplankton. The most significant types of phytoplankton include diatoms, cyanobacteria, and dinoflagellates, even though phytoplankton may take on a variety of shapes.

A great deal of phytoplankton is present, although it is not possible to perceive it with the human eye. Because of the existence of chlorophyll and auxiliary pigments within their cells (such as phycobiliproteins or xanthophylls), some species, when present in sufficient numbers, may display as colored patches on the surface of the water. This phenomenon is caused by the presence of chlorophyll (Domisch et al., 2015). This is the case for the simple reason that chlorophyll and auxiliary pigments are both green pigments. Chlorophyll is the primary green pigment in plants.

7.5.2 Kelp

Figure 7.7: Kelp.

Source: https://news.virginia.edu/sites/default/files/article_image/kelp_forest_douglas_klug_photo_lgdr.jpeg

Large brown algae such as kelps are members of the Laminariales order of algae. You may choose to explore through more than thirty different genera to narrow down your search. Despite its outward appearance, kelp is home to a wide variety of heterokonts, which are groupings of organisms that are not related to one another genetically. To this day, kelp has only been discovered in shallow seas that were home to so-called "underwater forests" during the Miocene epoch, which began between 5 and 23 million years ago.

The creatures thrive best at temperatures ranging from six to fourteen degrees Celsius (43 and 57 degrees Fahrenheit). Both Macrocystis and Nereocystis are well-known for their fast development, which allows them to reach lengths ranging from 30 to 80 meters in just a few days. This is a testament to their reputation (100 to 260 ft). In the nineteenth century, soda ash was more often known as "kelp," which was a term that applied to a wide variety of seaweeds that were burned for that purpose (primarily sodium carbonate). In the collection of algae that was used, members of the orders Laminariales and Fucales were discovered to be present in many species (Dar et al., 2020). During that period in history, the ashes after they have been cleansed were also referred to as "kelp."

If the water temperature is lower, there is a greater chance of finding kelp beds. Kelp beds like cooler temperatures. Kelp is a kind of algae that grows in water depths ranging from 20 to 90 feet near rocky shorelines. Kelp may be found in these water depths. It can do either because of its structure which resembles roots or its microscopic flotation bulbs, which give it the capacity to either cling to rocks or float on the surface of the water. It can grow to lengths of up to 76 meters and is light-dependent despite its status as the largest marine plant in the world (250 feet).

7.5.3 Red Algae or Rhodophyta

There are around 7,000 recognized species of Rhodophyta, and their taxonomy is constantly being re-evaluated as new information becomes available. The class Florideophyceae, which has the greatest number of species, has a total of 6,793 different kinds. The vast majority of these are multicellular marine algae, and among them are some of the most well-known types of seaweed. Red algae are found in substantially lower numbers in freshwater environments compared to their saltwater cousins. It is possible to find red algae in as much as five percent of all freshwater environments in the world in warmer parts of the earth, where the presence of this type of algae is more prevalent. There are no species that live on land, except two Cyanidiophyceae that live in caves. There is a possibility that an evolutionary bottleneck was responsible for the loss of around 25 percent of the capacity of our most recent common ancestor to adjust to different circumstances (Cardoso-Silva et al., 2018).

Figure 7.8: Red algae.

Source:https://cdn.shopify.com/s/files/1/0648/1955/files/Main_177328c7-a207-4c45-9120-79c1bd0bdc68.jpg?604326

Phycobiliproteins are a type of accessory pigment that is found in red algae and are responsible for the algae's characteristic red hue. Red algae are distinguished from other types of algae by the absence of flagella and centrioles in the eukaryotic cells that make up their bodies, the presence of chloroplasts that do not contain an external endoplasmic reticulum and unstacked (stroma) thylakoids, and the utilization of phycobiliproteins as accessory pigments.

Red algae store floridean starch outside of their plastids. Floridean starch is made up of amylopectin but not amylose. Red algae are the most prevalent form of red algae. They are big, multicellular, and found in saltwater. Red algae reproduce sexually and are the most common variety of red algae. In most cases, rather than having a life cycle that is defined by a two-generation cycle, the life cycle of the red algae is defined by a three-generation cycle. There is a chance that coralline algae can be found at this location. Coralline algae are responsible for the secretion of calcium carbonate, which is essential to the formation of coral reefs. Red algae, such as dulse and laver (nori/gim), are gathered for use in the food industry to make agar, carrageenans, and other components used in the manufacture of food products.

Around more than 500 million years ago, red algae began to dominate the world's warm and tropical seas and eventually took over the whole planet. There is no evidence that red algae may be found on almost every continent. The peculiar coloring is due to a pigment known as phycoerythrin, which absorbs blue light and then reflects it as red light. This is what causes the hue. This category of algae, which includes kelp, is loaded with a variety of nutrients, including protein, vitamins, and minerals (Benito et al., 2020). Because of this, a significant number of Asian countries depend on it as a primary source of food supply.

You will also find coralline algae, which is a type of algae that grows on rocks and is made up of filaments that are just as delicate as the rocks themselves. Coralline algae are essential to the growth of coral reefs because of their role in maintaining the structure of the reefs.

7.6 SEAGRASS

Figure 7.9: Seagrasses.

Source: https://images.theconversation.com/files/466144/original/file-20220531-24-oph25e.jpeg?ixlib=rb-1.1.0&rect=6%2C34%2C4594%2C3028&q=20&auto=format&w=320&fit=clip&dpr=2&usm=12&cs=strip

Other types of blooming plants can't be found in saltwater conditions for some reason. The order Alismatales is comprised of four families of seagrasses, each of which is unique to the marine environment and can only be found there. Among the many plant families that are represented here are the Posidoniaceae, Zosteraceae, Hydrocharitaceae, and Cymodoceaceae (in the clade of monocotyledons). There was a recolonization of the oceans by land plants between 70 and 100 million years ago, which led to the creation of seagrasses. Some plants have an appearance that is comparable to that of grasses belonging to the Poaceae family.

Seagrasses and other plants that can sustain themselves through photosynthesis are nourished by sunlight that penetrates the water's surface. They are at their most active in waters around the coast that are warm, shallow, and sheltered, and that have sandy or muddy bottoms. Because the majority of marine species finish their entire cycles underwater, the process of underwater pollination happens rather frequently. Pollination of *Thalassia testudinum*, a species that was traditionally thought to be pollinated entirely by the movement of sea currents, has been shown to take place via a mix of both biotic and abiotic mechanisms (Bassi et al., 2014). This contradicts the traditional belief that pollination takes place entirely due to the movement of sea currents. Contrary to what was previously assumed, pollination cannot take place only by this approach. Crabs, *Majid zoae*, and *Thalassinidea zoae* are only some of the crustaceans that have been discovered to carry pollen grains. This suggests that the plant develops nutrient-rich mucigenous clusters of pollen to attract and cling to the crustaceans, rather than nectar-like terrestrial flowers do.

Underwater seagrass meadows are among the most numerous ecosystems on Earth since seagrass is the primary component in their formation. Seagrass meadows are home to a diverse array of creatures, including both plants and animals, making them an important habitat. In addition to acting as vital carbon storage and supplying homes and food for a wide variety of marine organisms, they also perform the function of coral reefs, although they are not composed of coral.

Seagrass is a rooted plant that can be found growing anywhere in the world and can be found in the shallower parts of the ocean. There are many different kinds of seagrass, but the most common ones include manatee grass, turtle grass, and shoal grass. All of the land plants that have evolved to be able to live in salty water are descended from a single progenitor (Balarabe, 2006).

When temperatures are moderate, it flourishes in the marshy waters where crabs and lobsters may be found, and they rely on it as a source of nourishment. Blooms of seagrass produce pollen that helps fertilize other ocean plants and also oxygenate the water. This doubles as a benefit to the ecosystem as a whole. It provides a home for the marine life that lives there as well as a source of food for the many different kinds of reef fish. As a result, seagrass has become an important component of marine ecosystems all over the world.

7.7 SARGASSUM

Figure 7.10: Sargassum.

Source: https://sc04.alicdn.com/kf/Ud4ccfd9257cf458e8bbb34d6c-408d7aeC/945689078/Ud4ccfd9257cf458e8bbb34d6c408d7aeC.jpg

Brown sargassum is a kind of macroalgae (seaweed) that is classified under the Fucales order of algae (class Phaeophyceae). The capacity of species of the genus that live in plankton to float freely in water is one of the genus' most recognizable characteristics. In waters across the world that are either temperate or tropical, the ideal sites to seek these species are areas with shallow water and coral reefs. The majority of Phaeophyceae species are found in cold water because upwelling causes nutrients to rise to the surface. Sargassum, on the other hand, seems to buck this general trend in a significant way. There are several circumstances in which benthic creatures that have been driven from their reef habitats as a result of harsh weather events may make the shift to either a planktonic or a pelagic way of life. Despite this, two of the species, *S. natans* and *S. fluitans*, have become holopelagic, meaning that they reproduce vegetatively and do not attach themselves to the bottom of the ocean. As a result of the high concentration of algae in the Atlantic Ocean, the Sargasso Sea has made its home there (Barbosa et al., 2012). In addition to helping the plant produce more oxygen through photosynthesis, the capacity of the plant's bladders to float makes it possible for juvenile sea turtles to rely on the plant for both food and shelter as they mature and get ready to head back to their breeding grounds.

CHAPTER 8

INLAND WATER RESOURCES EXPLOITATION (FISHING, MINING, POWER GENERATION, TRANSPORTATION)

CONTENTS

8.1 INTRODUCTION

Figure 8.1: Inland water resources.

Source: By M M from Switzerland - Amazonas, Iquitos - Leticia, Kolumbien, CC BY-SA 2.0, https://commons.wikimedia.org/w/index.php?curid=40537857

The ecosystems and species that live in inland water are in a less favorable situation (with medium confidence), in comparison to those that live in coastal, grassland, and forest habitats (medium certainty). It is estimated that throughout the twentieth century, inland water habitats suffered a loss equivalent to fifty percent of their original number. It is common knowledge that the ability of inland water systems to provide a wide variety of ecosystem services is dwindling and is on par with or possibly falling behind that of other types of water systems. This is the case even though this capacity has been known for quite some time (Abubakar & Abubakar, 2013). There is not enough evidence to support the widespread nature of habitat loss across the globe, particularly concerning habitats that are dependent on seasonally changing or intermittent water sources. In certain parts of North America, Europe, and Australia, more than half of the inland water has evaporated during human history.

As a direct consequence of this, the ecosystems of inland waters have been decimated. In addition to the disappearance of inland water systems, there has also been a significant amount of deterioration. Despite the extensive documentation of various site-specific occurrences, it is not possible to construct an estimate for the entire continent with a high degree of certainty. In a similar vein, it has been challenging to arrive at an accurate estimate of the amount of habitat that has been lost. The loss of species variety in the world's inland water settings is occurring at a startlingly rapid rate, placing these areas among those that are under the greatest danger. As a result of changes in the global climate, it is anticipated that a great number of inland water systems, in addition to species, will suffer losses and deterioration. Despite this, estimates do not yet have clear facts to show the degree to which such losses, degradation, or decline will occur.

Whether they have done it alone, in cooperation with one another, or jointly, several circumstances have contributed to the deterioration and destruction of inland water systems. This is true whether they have done it alone, in cooperation with one another, or jointly. For instance, changes in land use or cover caused by the removal of vegetation; drainage and infilling, particularly as a result of agricultural development; the spread of infrastructure; the introduction of exotic species; hydrologic adjustment; and overharvesting via fishing are all primary sources of deterioration and loss of inland water (Alcocer et al., 2021). Alterations in land use or cover as a result of the clearing of vegetation, drainage, and filling, most notably as a result of agricultural development; the introduction and establishment of new uses and covers; and even though many stressors have accumulative and synergistic effects, there is no sufficient quantitative evidence to show their relative importance or the proportionate individual and combined repercussions of those stressors.

There is a high degree of certainty that the expansion of agriculture throughout human history has been the primary factor in the widespread depletion of inland water resources (high certainty). By the year 1985, intensive agriculture had resulted in the draining of between 56 and 66 percent of the exploitable inland water systems in Europe and North America, 27 percent of those systems in Asia, and 6 percent of those systems in South America. Dams and other riverside structures have modified or regulated the flow of approximately sixty percent of the world's major river systems. This accounts for approximately forty percent of the world's major river systems. To put it another way, it constitutes almost two-thirds of the whole. The construction of large dams continues to be challenging in many different countries. Because of eutrophication and other forms of water pollution, an uncountable number of inland rivers have been destroyed in a number of different countries.

Because of pollution, the ability of inland streams to filter and digest waste has been significantly reduced, which has a negative impact on the biodiversity of the region. In regions with limited water supplies, concerns regarding water quality have to be addressed more quickly (through the use of dryland systems). Never before have so many people and ecosystems been put in danger by the combination of dangerous chemicals and compounds made by humans. It is difficult to speculate on the long-term effects that this will have on ecosystems as well as on human health (Wu et al., 2021). In recent years, the havoc that invasive species wreak across the globe has received an increasing amount of attention as a result of increased awareness of the problem. As a result of pollution and degradation, the positive effects that inland waterways have on both the natural world and on people's lives have been diminished.

Because more than fifty million people are directly involved in inland fisheries, it is essential for the upkeep of human wellbeing that services such as fish be delivered from inland waterways. More than half of the world's population considers rice to be an essential part of their diet, and it is anticipated that by the year 2020, this number will have increased to 4 billion people. The preservation of ecological processes and the exercise of control over them are of the utmost importance in ensuring their continued existence over time. Flow management both within and across inland rivers, as well as linkages between surface water and groundwater, are examples of biological services that are failing on a global scale. Another example is the management of surface water quality. The typical flooding patterns of a great number of river systems have been disrupted, which has led to a reduction in the movement of sediment, a loss of flood buffering, and an increase in the retention of nutrients.

However, floods such as the one that occurred in China in 1998, which caused an estimated $20 billion in damages, are necessary for the maintenance of healthy floodplains and the sustenance of fish populations in important rivers. Inland rivers not only provide essential opportunities for recreation and tourism to several different cities and towns, but they also have significant value on a variety of other levels, such as those of aesthetics, art, education, culture, and spirituality. Inland rivers have a significant impact on all of these aspects of life (Tundisi et al., 2010). However, in the majority of countries, the economic value of these services for many local ecosystems is typically not accurately measured, is not recognized by politicians, and is not prioritized by resource development and conservation agencies. This is in spite of the fact that the importance of these services to the economy is widely recognized by everyone. In recent years, the enormous but largely underappreciated trade-offs that are present between the benefits provided by inland waterways have become more apparent.

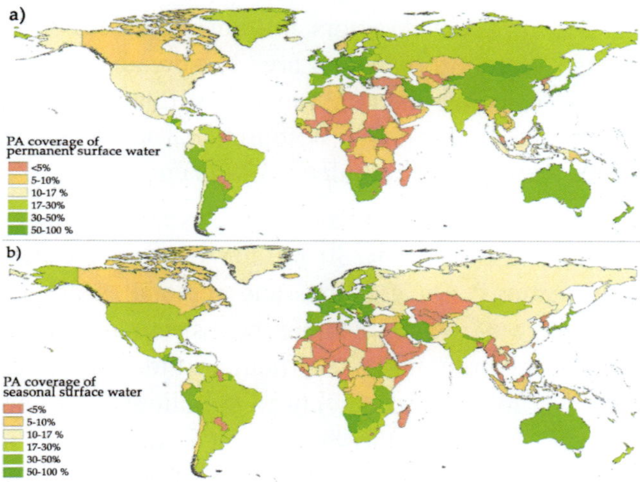

Figure 8.2: Inland water sources.

Source: https://journals.plos.org/plosone/article/figure/image?size=inline&id=10.1371/journal.pone.0210496.g005

The river infrastructure has increased agricultural productivity by making more land and irrigation water accessible, in addition to its positive effects on transportation and the control of flooding, and the generation of power. Because of the river's infrastructure, these improvements are now within reach and can be implemented. As a result of rivers being redirected away from their floodplains and other inland water habitats in certain locations, riverine water velocity has decreased; rivers have been transformed into a series of lakes that are connected, and the rate at which groundwater is recharged has decreased.

The construction of infrastructure has increased the potential for flooding in other parts of the country because it channels water and speeds up flow rates. This has led to an increase in the number of flood-related deaths. These changes have resulted in a loss of freshwater biodiversity and inland fishing options, as well as altered fish migratory patterns and altered the nature of the riparian habitat. In addition, these changes have caused a loss of riparian habitat. Additionally, as a result of these modifications, coastal ecosystems have been altered, and new passageways for invasive species have been established (Togaki & Katano, 2020). Waterlogging and pollution, especially eutrophication and salinization, not only cause damage to the irrigation system and other functions but also speed up the process of introducing human disease vectors into an area. This makes it easier for human disease vectors to enter an area or spread throughout an area. It is not possible to calculate the total number of inland water systems and the progression of these systems on a continental scale as a result of the imprecision of the data and the turbulence of the environment.

The total area of the world's inland water systems is expected to range anywhere from 530 million to 1,280 million hectares by the year 2050, according to projections. On a global and regional scale, the size and distribution of inland waterways are either inconsistent or poorly understood due to definitional ambiguity and the difficulty in identifying and mapping ecosystems with variable borders owing to variations in water level. This difficulty is largely caused by the fact that it is difficult to identify and map ecosystems with variable borders because of changes in water levels. When considering issues on a global scale, this point becomes especially important. In addition to significant rivers, larger wetland regions, lakes, and inland seas have been reported; there are approximately 10,000 lakes that are larger than 1 square kilometer, and peatlands comprise more than 400 million hectares.

In the United States alone, the number of significant rivers is estimated to be in the tens of thousands. It has been reported that a significant portion of the continent is covered by inland seas. However, despite the importance of smaller habitats to a wide variety of species, these areas have not been sufficiently explored or designated. The vast majority of the data that is currently accessible focuses on larger scales that are either regional or global in scope. This makes several research activities more challenging, and caution is required when attempting to compare multiple data sets, particularly when the data sets were gathered at different geographic sizes (Teffera et al., 2019). Especially in regions where data is scarce, insufficient, or out of date, there is a need for creative ways to analyze the state of inland water systems and the patterns associated with the species that live there.

This is especially true for the patterns associated with the species that live there.

The amount of water that is both available and used is impacted by several factors, including the physical geography, climate, and size of a country's catchment area, as well as the availability of water resources and the level of socioeconomic development. There is a significant amount of variety in the pattern of precipitation that the region experienced as a result of factors such as geography and the passage of time. The western slopes of the Indian and Indo-mountains, as well as the southern slopes of the Himalayas, China, and the Indonesian islands, receive average annual precipitation of 1,500 millimeters or more than 3,000 millimeters. This range of precipitation is common throughout these regions. On the other hand, on an annual basis, the northern part of the region receives less than 200 millimeters of precipitation on average. During the protracted dry season, many river basins are subjected to recurrent water shortages, whereas flooding is a common occurrence during the wet season.

The cyclical monsoon rainfall pattern is the most common type of rainfall pattern in the countries that border the Indian and Pacific Oceans. This type of rainfall pattern is characterized by distinct dry and wet seasons. Asia and the Pacific are responsible for more than a third of the total annual runoff produced throughout the world, which amounts to approximately 13,260 km3 in volume. Even though many of the less developed countries in the region have significant amounts of annually renewable water resources, not all of them are easily accessible for human use (Soranno et al., 2010). The People's Republic of China, Indonesia, and Pakistan are the countries that have the most water resources. However, when the data is broken down population by population, a very different picture emerges.

The disparity in the region's annual per capita internal (within a nation's borders) renewable resources is illustrated in figure 4.2. In Southeast Asia, the amount of annual renewable water resources that are available per person ranges from slightly more than 172 cubic meters in Singapore to more than 21,000 cubic meters in Malaysia. Annual per capita internal (within a nation's borders) renewable resources. At this time, Singapore obtains a portion of its freshwater supply from other countries, specifically the People's Republic of China and Malaysia. The freshwater resources of South Asia, which are comprised of the countries of India, the Islamic Republic of Iran, the Republic of Korea, and Pakistan, range from 1,400 to 1,900 m3 per capita per year, which is a quantity that is significantly lower than that of Malaysia.

8.2 FISHING EXPLOITATIONS

Figure 8.3: Fishing exploitation.

Source:http://www.aquaticlifelab.eu/wp-content/uploads/2018/06/Photo-4.3.1..jpg

Fish are a critical source of food for billions of people across the world, and for millions more, fishing is a lucrative profession. Seas and seas have long been seen as a never-ending source of sustenance. However, throughout the past half-century, fishing demands and unethical tactics have driven countless fish species to the edge of extinction. Overfishing arises when a population's fish stocks are diminished faster than natural reproduction can replace them. It has been estimated that 75 percent of all or abundantly fished fish species worldwide have been depleted by human activities. Fisheries in more than 30 percent of the world are overexploited and need rigorous control to rehabilitate them. It is now likely that some of the world's major commercial fish species, such as Atlantic Bluefin tuna, will become extinct. Tuna and grouper stocks are being harmed by overfishing for top predators, which has led to a surplus of lesser marine species like sardines and anchovies (Sigala et al., 2017). Fish stocks in the Mediterranean are under significant strain due to economic growth, population rise, and tourism, as well as decades of overfishing. Fish populations are at risk of depletion or exhaustion when clandestine fishing, discards, pollution, and deficiencies in data collection and maintenance are taken into account.

Overfishing threatens at least 96 percent of bottom species, but pelagic fish such as sardines and anchovies are at risk at a rate of 71 percent or higher, according to research. Some fish, such as hake, red mullet, black-bellied anglerfish, and blue whiting, are dying at six times the acceptable rate, according to scientists. Since 1950, the Mediterranean has lost 41 percent of its marine species and 34 percent of its fish populations. If the current level of stress on the Mediterranean area is not alleviated immediately, the ecosystem may be driven to the point of no return. There are an increasing number of fishing communities that now recognize the significance of mending the maritime environment to rehabilitate fish populations, raise catch, and enhance their quality of life. Fishing reforms must go hand in hand with major advances in other maritime enterprises, such as tourism, which is largely dependent on the healthy Mediterranean. This trade is putting beautiful creatures in jeopardy because buyers aren't educated enough about what they're buying. In the long term, regulating purchasing power may make a major influence on the usage of food resources. Fishermen can secure their resources and apply procedures that have a low environmental effect thanks to solid scientific data collection.

An excessive amount of fish is being harvested out of the sea by overfishing. For a multitude of factors, such as "chronic overcapacity" of current fishing fleets to take far more fish than can be replenished (which is the most major reason), overfishing arises. During the years 1975-1995, the worldwide catch of marine fish more than doubled, and by 1999, the global fisheries situation had reached crisis proportions. Fishing communities, governments, and non-governmental organizations are all anxious about catching too many undesirable species (Simões et al., 2011). Fisheries may be more profitable if they capture every fish that can be taken and invest the revenues. Restrictions on catch may cause overfishing of some species when they are introduced. There were times when it was more cost beneficial to take any blue whale that happened to pop up, even if the whaler was primarily targeting the more valuable fin and humpback whales because quotas were defined in terms of Blue Whale Units. More frequently than not, overfishing is placed at hazard by the intricacy of developing international conservation standards.

Academics say that only a coordinated effort can avert the collapse of the world's major fisheries. Marine reserves should be developed swiftly, fishing fleets downsized, and subsidies are withdrawn, to the tune of $2.5 billion per year in the North Atlantic region alone in 2001.

Figure 8.4: Fish exploitation.

Source: https://ychef.files.bbci.co.uk/976x549/p00ytqyb.jpg

More than half of the world's fisheries are overexploited, depleted, or recovered from depletion, according to the WWF 2014. The yearly fish collection rose until it reached a record high of 100 million metric tons in 1989, expanding fivefold since 1950 and doubling since 1963. Worldwide fish capture has been stable at about 97 million tonnes since 1989. Many persons believe that international catches have declined in recent years but they claim that earlier tonnage growth disguised a transition from marketable species to substantially less desired ones. As a result, the percentage of impoverished nations has reduced. This was largely attributed to a reduction in the amount of microscopic pelagic fish collected in the southeastern Pacific Ocean, which had been responsible for an increase in 1986. For example, from 52.7 percent to 51.5 percent in 1987. Small pelagic fish gathered in the southern Pacific dropped by 51.5 percent in 1987, a fall from 52.7 percent in 1986 and an increase in that year.

The other four had already been tired of their vitality by this point. In the North Atlantic Ocean, fish populations in 2000 were less than one-sixth of what they had been in 1900. This conclusion was significantly worse than earlier estimates had projected. The decline in catches made it apparent that something was incorrect. The amount of prime table fish taken per person has fallen from 9 kilos in the 1960s to 3 kilograms in 2000. If the current trend continues, the majority of fish will be unavailable for purchase by the year 2010. Fishermen in Western Europe and the United States were able to "conceal" the problem of a North Atlantic fish scarcity by moving more and more fish from other regions of the world to these destinations (Shah et al., 2019). There were losses of 14, 41, and 23 percent correspondingly in the output of fish in Chile, Ecuador, and Peru between the peaks of 16 million tonnes in 1986 to the end of 1988.

There has been a decline in the fishing of 9 of the 12 Atlantic groundfish populations. More valuable fish like cod, haddock, and flounder have suffered a 75 percent drop in capture. Western Atlantic bluefin tuna and Pacific salmon both face extinction because of overfishing and habitat loss. The spawning stock of cod in the North Atlantic declined from 264,000 tonnes in 1970 to fewer than 60,000 tonnes in 1995, according to estimates. Overfishing of immature North Atlantic swordfish decreased their numbers by 70 percent. Species extinction is a concern in nearly all of the commercial fishing zones along the US coast. The Georges Bank region, a longtime fishing hotspot in the United States, has been stripped bare in the previous decade and is now used only for fishing. As with the cod population, the spawning stock on Canada's Grand Banks has decreased from 1.6 million to 22,000 individuals. Stock collapse was predicted to constitute a threat to more than half of the fisheries around the UK in 1996, when the spawning population was at its peak. Stocks of plaice in the North Sea decreased by 49% from 1986 to 1996. As recently as 1996, there was only a fraction of the cod stocks there were in the 1980s.

Although EU/EC fishing permits are periodically reduced, the North Sea cod spawning stock was less than 80,000 tons in 1990, down from 168,000 tons in 1982, notwithstanding this. During the same timeframe, haddock stocks fell from 285,000 to 80,000 tonnes. EU/EEC agreed in 1992 to reduce its fishing fleet by at least 20 percent by the end of the decade.. Despite the fact that cod accounts for about a third of all Icelandic exports, it has banned cod fishing off the Canadian coasts, and Iceland has reduced its cod catch by 40%. In order to enforce quotas, Namibia and Sierra Leone have both banned foreign vessels from their seas. Eight island states in the South Pacific Forum Fisheries Agency have agreed to limit the Asian and American fishing fleets to 200 vessels each, despite serious concerns (Simmatis et al., 2021). Over the UN driftnet ban, the United States and China; Ecuador, between a developing national fishing industry and small-scale traditional fishermen; the Caribbean, with illegal fishing by Japan; Russia; Taiwan; the United States; and Venezuela; the Patagonian Shelf; the United States and France over Bluefin tuna, and finally the Bay of Biscay.

8.3 MINING EXPLOITATION (EXPLOITATION OF MINERAL RESOURCES)

Figure 8.5: Mining exploitation.

Source:https://images.pexels.com/photos/461789/pexels-photo-461789.jpeg?auto=compress&cs=tinysrgb&dpr=1&w=500

Antimony is one of the mineral resources that the general public is unfamiliar with. Antinomy is commonly found in batteries, ceramics, electrical wire sheathing, enamels, fire-proofing additives, glass, metal alloys, pigments, and semiconductors. The ancient Egyptians first used antimony in the formulation of cosmetics and medicinal compounds. Its use as a component of alloys became popular in the 1500s. Antimony alloys were used in a variety of industries by the late 1700s, including steam engines and bullet production. According to the United States Geological Survey, China produces the vast majority of the world's mined antimony. Although antimony is found in soils all over the world, the majority of extractable antimony resources are found in China and Russia. It is expected that the world's antimony reserves will be depleted by 2060.

Bismuth is a less well-known mineral than antimony, but it can be found in all soils. Bismuth deposits can be extracted in two forms: bismuthinite, which is a sulfur-rich compound, and bismite, which is a bismuth oxide-rich compound. People believed that these were made of lead from ancient times until the 1400s. Although it is possible to mine bismuth on its own, it is usually obtained as a byproduct of the lead mining industry (Robarts & Zohary, 2018). The mineral was previously used in medical preparations and colorants, but it has recently found new applications in alloys and electronic devices.

Boron is well-known for its industrial applications; however, a growing body of research supports its use in medicine, as discussed in Boron-Base Compounds by Evamarie Hey-Hawkins and Clara Teixidor. Boron is a chemical element represented by the symbol "B." Since around 300 CE, the Chinese and Persians have used boron as an ingredient in ceramic glazing; today, more than half of the boron mined globally is used in the production of various glass and ceramic items. Boron is also found in semiconductors, fertilizers, and household cleaning products. Even though deposits can be discovered in any type of soil, they are most commonly found in China, Russia, Turkey, and the United States. Even though Turkey holds the vast majority of the world's boron reserves, the United States of America is a significant boron exporter. The current levels of global boron consumption, according to the United States Geological Survey (USGS), will not result in depletion for another 250 to 500 years. The cost of extracting the boron is a more pressing concern than the resource's depletion.

According to Robert Krebs' research, published in The History and Use of Our Earth's Chemical Elements, chromium is less common in soil than other metals and can only be mined as chromite ore. The three countries with the most are South Africa, India, and Kazakhstan. The extraction of pure chromium from its ore takes a long time and adds to the overall cost. Swapan Haldar's book Platinum-Nickel-Chromium Deposits: Geology, Exploration, and Reserve Base is an excellent resource for learning more about chromium's unique properties and practical applications. Chromium is an extremely important mineral that is used to make a variety of products such as chrome plating, colors, and stainless steel. It was discovered in 1797 in France, but it did not have any significant economic applications until the early 1900s (Ribas et al., 2019). The mining of chromium increased around 1950, and this trend continued. It is expected that chromium mining will begin to decline in 2030 and will reach depletion by 2100.

Surface mining techniques have been used by humans to extract copper ore since around 9000 BCE. It was refined into a pure metal around 5000 BC, and during the Copper Age, it surpassed the use of many nonmetallic minerals (3000 BCE). Furthermore, it was refined into a pure metal around

5000 BC (3000 BCE). Copper is an essential metal that is used in a variety of modern electrical devices, transportation equipment, and building materials, according to author Günter Joseph's book Copper. Copper consumption is expected to rise as these three economic sectors expand around the world. Copper deposits can be found all over the world, but the largest ones are in Australia, Chile, Mexico, and Peru. Copper is most commonly found in products made in China and the United States. According to projections, copper reserves will begin to decline steadily by 2050 and will be completely depleted by 2100.

According to Giulio Morteani and Jeremy Northover's book Prehistoric Gold in Europe, the earliest evidence of human use of gold dates back to 40,000 BCE; however, historians trace its prominence as a treasured resource back to Egypt around 3000 BCE. Even though small amounts of gold may be found in other parts of the world, South Africa is home to the majority of the world's gold reserves. In terms of gold mining output, Australia, China, Canada, Peru, Russia, and the United States have recently surpassed South Africa. China currently produces the most gold of any country on the planet. Despite its many other uses, gold is highly valued in the jewelry industry and monetary transactions. According to Pierre Vilar's book "A History of Gold and Money: 1450–1920," gold, along with iron, is widely recognized as having the greatest historical significance. Gold mining is expected to increase until 2050 when it will begin a precipitous decline that will last until 2060 when it will end.

Rhenium is a transition metal element on the periodic table with a name that sounds more like a disease. Even though its existence had been hypothesized as early as 1912, rhenium was not discovered in platinum ore until 1925. According to J. G. F. Druce's Rhenium, it's not very abundant in soil, but it's abundant in a shiny rock called columbite. According to the United States Geological Survey (USGS), the majority of rhenium comes from copper mining waste. Rhenium is used in the production of lead-free gasoline, as well as the production of lightweight metals (Reed et al., 2008). The two countries that produce the most rhenium are the United States of America and Chile, respectively. It is expected that rhenium mining will peak in the year 2100, then rapidly decline until it is exhausted in the year 2200.

Tin has been used to make bronze since 3500 BC; the Bronze Age began due to the metal's versatility, which allowed it to be used for a wide range of purposes. Tin has been an essential component in a variety of fields, including the defense industry, food preservation industry, and technological device industry, according to the book Tin and Global Capitalism, 1850–2000 edited by Mats Ingulstad, Andrew Perchard, and Espen Storli. The development of superconducting materials has resulted in a slew of novel new applications. Although China, Indonesia, and Peru produce the most tin, Southeast Asia and China have the majority of the world's tin resources. Tin output is predicted to rise until 2025 when it will begin a gradual decrease that will last until the year 2100 when it will reach depletion.

Figure 8.6: Mining.

Source:https://media.istockphoto.com/photos/large-open-pit-mine-for-excavation-and-exploitation-picture-id1364787448?k=20&m=1364787448&s=612x612&w=0&h=aWmYDq3Zt15E2R8ECCbNfp11v04zqmeObX77Gafr8II=

According to Robert Lauf's Mineralogy of Uranium and Thorium, uranium is a common component of a variety of soils and may even be found in seawater. Uranium ore comes in several isotopes, often known as forms. Because uranium-235 is used in nuclear fission reactors, it is the most valuable of all uranium isotopes. It was first used as a pigment in ceramics and glass in the years following its discovery in Germany in 1789; its radioactive qualities were not identified until 1896. The prospect of uranium being utilized as a cancer treatment sparked a boom in the uranium mining business in the early 1900s. After scientists gained control of the nuclear fission processes, uranium became a more valuable resource (Pinilla & Gabriel, 2016). This has the long-term consequence of irreversibly changing the dynamics of energy production and conflict resolution. Kazakhstan, Australia, Russia, Canada, and Niger have the greatest reserves of high-grade or high-concentration uranium. In the United States, uranium is only available in restricted amounts. It is expected that worldwide uranium reserves will begin to diminish around 2040 and will be completely exhausted by 2100.

Zinc may be found in very high quantities in sphalerite ore deposits, although it is only present at minimal levels in many soils. Zinc was found circa 3000 BCE in the Middle East and was later mined for its usage as a component of brass. Brass was determined to be an ideal alloy due to its malleability and corrosion resistance, according to the information offered in Understanding Copper Alloys by Howard Mendenhall. Furthermore, it may be found in a wide range of commercial and consumer items. Zinc mining occurs all over the world due to the fact that zinc may be found in a variety of soil types. At the moment, the majority of zinc is taken from the earth in China and Australia. Zinc extraction is expected to expand continuously until around 2040 when it will begin to drop steadily and continue on this path until it hits depletion around the year 2100.

Even the most abundant natural resources, such as aluminum, feldspar, and silica, face the possibility of depletion shortly. According to the authors of Metals and Society, Nicholas Arndt, Stephen Kesler, and Clément Ganino, bauxite and aluminum are common minerals found in the majority of soils. Feldspar, the non-metallic mineral, accounts for around half of the Earth's crust, according to Feldspar editors Francis Whitmore and Jorge Escamilla. The author of the Society for Mining, Metallurgy, and Exploration book Crystalline Silica states that silica may be found in practically every type of rocky material and that its refined form is used in the creation of glass and hardening agents. Even though there are still substantial amounts of untreated rocks that may be mined, the concentrations of these three minerals are progressively falling. Untreated resources have the potential to be used in the manufacture of building materials as well as road construction (Pinilla, 2010). Rock quarries may be found in almost every part of the world, and their numbers are growing. The separation of pure aluminum, feldspar, and silica is also becoming an increasingly crucial phase in today's contemporary mining processes. According to the United States Geological Survey, quarry rock will be easily available until the year 2050. However, in the year 2100, the accessible supply of aggregate rock may begin to decline dramatically.

8.4 THE EXPLOITATION OF POWER GENERATING RESOURCES

Figure 8.7: The exploitation of power generating resources.

Source: https://thumbs.dreamstime.com/b/wood-energy-icon-linear-black-rgb-color-styles-power-manufacturing-business-electricity-generation-natural-resources-192710338.jpg

In recent years, there has been a greater emphasis placed on the use of renewable energy sources as a source of power generation. As a result, economic concerns and attempts to ensure energy supply dependability are being undertaken. A large rise in energy consumption is required to support economic growth. These three resources are no longer adequate to fulfill the demand for carbon, natural gas, and petroleum, in that order. Traditional energy sources are a major contributor to the deterioration of environmental conditions and the acceleration of climate change.

When additional renewable energy sources are incorporated into a country's infrastructure, it guarantees that its people have easy access to local natural resources. In times of crisis caused by the depletion of energy supplies, as well as other political or economic issues, the nation becomes self-sufficient due to the increased resources available for energy production (Pérez-Fuentetaja et al., 2006).

Renewable energy is playing an increasingly important role in the global energy supply. These changes are being pushed by laws, evolving energy technologies, and a growing public concern for the well-being of wildlife and the natural environment.

We have begun to experiment with cutting-edge technologies such as the conversion of biomass into fuel and the usage of distributed power generating systems because we are concerned about our country's energy supply. Based on current trends, it is expected that the generation of renewable energy would be the industry that expands at the fastest pace in rural regions.

Figure 8.8: Power generators.

Source:https://www.bioenergyconsult.com/wp-content/uploads/2018/11/UK-renewable-energy-supply.jpg

Because supplies of oil, natural gas, and nuclear fuel are all running low, there has been a renewed emphasis on developing alternative energy sources that are less harmful to the environment. During the 1970s, the dramatic fall in oil supplies from the Middle East to some of the world's most industrialized nations drew attention to the volatility and unpredictability of petroleum fuel availability. It has also been discovered that our reliance on fossil fuels harms the environment. The principal sources of sulfur dioxide and nitrogen oxide emissions into the environment are combustion engines and steam turbine power plants that use coal or natural gas as fuel. The final compound is very acidic because sulfuric and nitric acids were mixed with water vapor during the precipitation process (Papa & Mamaril, 2011). The burning of fossil fuels produces carbon dioxide in addition to contributing to global warming.

Carbon dioxide concentrations in the atmosphere have steadily increased since the mid-nineteenth century. This rise is mostly due to the increased usage of fossil fuels such as coal, oil, and natural gas. Scientists are becoming increasingly concerned about the prospect of a greenhouse effect, which would result in a warming of the Earth's surface by increasing the amount of heat trapped in the lower atmosphere as carbon dioxide levels in the sky rise (along with other industrial gases like methane and chlorofluorocarbons). Climate changes caused by this illness may have a considerable detrimental influence on agricultural operations and natural environments. Read the articles "Global Warming," "Climatic Variations and Change," and "Hydrosphere: Acid Rain and Greenhouse Gas Buildup" for a more in-depth examination of the relationship between acid rain and the greenhouse effect. Governments throughout the world are investing in alternative energy research and development to reduce their dependency on fossil fuels. Many people feel that fusion devices provide the most potential long-term answer since they draw the majority of their power from the plentiful isotope of hydrogen known as deuterium. The sun, wind, water, and geothermal energy are all potential sources of energy that might be harnessed with other technologies that are now being developed (i.e., geothermal energy). Wind and solar power are two examples of renewable energy sources that cause nearly no pollution; nevertheless, only a tiny part of the energy produced by these sources can be transformed into useable electricity at a reasonable price (Pathak & Mudgal, 2005).

8.5 THE EXPLOITATION OF TRANSPORT RESOURCES

Figure 8.9: The exploitation of transport resources.

Source: https://www.umweltbundesamt.de/sites/default/files/medien/378/bilder/rohrleitung_im_bau_reinhardt_fotolia_46815756_m.jpg

Since supplies of oil, natural gas, and nuclear fuel are all diminishing, there has been a heightened emphasis placed on discovering alternative energy sources that are less destructive to the environment. The dramatic decline in oil supplies from the Middle East to some of the most industrialized nations in the world throughout the 1970s brought attention to the volatility and unpredictability of the supply of petroleum fuel. It has also come to light that our reliance on fossil fuels harms the surrounding ecosystem. Combustion engines and steam turbine power plants that utilize coal or natural gas as a fuel are the primary sources of the emissions of sulfur dioxide and nitrogen oxides into the environment. Because sulfuric and nitric acids were combined with water vapor during the precipitation process, the resulting substance is highly acidic. In addition to contributing to global warming, the combustion of fossil fuels results in the production of carbon dioxide.

Since the middle of the 19th century, there has been a steady increase in the concentration of carbon dioxide in the atmosphere. This increase is mostly attributable to the increasing consumption of fossil fuels like coal, oil, and natural gas. As a result of rising levels of carbon dioxide in the atmosphere, scientists are growing increasingly concerned about the possibility of a greenhouse effect, which would result in a warming of the Earth's surface by increasing the amount of heat that is retained in the lower atmosphere (along with other industrial gases like methane and chlorofluorocarbons). Changes in climate brought on by this illness might have a significant negative influence on agricultural practices and natural environments. See the entries global warming, climatic fluctuations and change, and hydrosphere: Acid rain and the Buildup of Greenhouse Gases for a more in-depth explanation of acid rain and the greenhouse effect (Owusu-Boateng et al., 2022). It is envisaged that renewable energy technologies would minimize the countries' dependency on fossil fuels and the attendant issues.

Since fusion devices derive the majority of their power from the abundant isotope of hydrogen known as deuterium, many people believe that they provide the most promising long-term solution. The sun, the wind, the water, and geothermal energy are all potential sources of energy that might be harnessed via the use of other technologies that are now in the process of being developed (i.e., geothermal energy). Wind and solar power are two examples of renewable energy sources that produce almost no pollution, but only a small part of the energy that these sources produce can be converted into usable electricity at an affordable price.

Typically, mineral and petroleum resources are found on land, and transportation networks have facilitated their utilization. In the United States, iron ore and coal are two of the most regularly traded commodities. When it comes to maritime shipping, petroleum is by far the most significant cargo. Most petroleum is consumed by automobiles, with highway vehicles accounting for more than half of the overall consumption. There has been an increased focus on car fuel efficiency since the early 1970s when oil prices began to climb. There is a clear association between dropping fuel costs and a growth in the number of individuals driving their automobiles rather than utilizing public transit.

Even though the negative environmental consequences of transportation infrastructure building have been considerably decreased throughout time, the construction of these structures was fundamentally damaging to the environment. When it comes to large-scale measures to avoid interfering with the migration patterns of particular animal species, the Trans-Alaska Pipeline is the biggest example. Noise control has been enhanced through the erection of sound-deflecting barriers near metropolitan freeways and insulating buildings near airports. The government is phasing out the utilization of aviation engines that create greater noise. The removal of obsolete and abandoned vehicles has spawned a new land-use dilemma. Some automobiles are abandoned on city streets or in fields, and junkyards are unattractive (Ohtaka et al., 2011). The materials used to make autos restrict recyclers from paying a high price for junked automobiles to be transformed into scrap for resale and reuse.

Figure 8.10: Means of transport.

*Source:https://miro.medium.com/max/505/1*ltYV108kfzhASGi2sHJWPQ.png*

Additionally, transportation has affected the water supply. Near port sites, marshes have been filled in for millennia to allow cargo handling and industrial infrastructure. Important rivers have been controlled through the placement of dams and locks. Before the introduction of sewage treatment facilities, urban and industrial pollutants were carried through rivers and streams. To eliminate accumulated silt, harbors and navigation channels must be dredged regularly. This dredged material is occasionally unclean, and there are arguments over where it should be deposited to minimize environmental impact. The loading and unloading of dry bulk cargo cause dust, necessitating the development of complicated dust collection systems by port facilities. Every year or so, a large oil spill occurs somewhere in the world. However, oil spills are generally controlled before a huge tragedy strikes.

CHAPTER 9

GLOBAL CRISIS IN THE FRESHWATER RESOURCES

CONTENTS

9.1 INTRODUCTION

Water is a natural resource that is readily available for people and animals across the world. It is considered a foundation for livelihoods and life and is a key ingredient in sustaining development. As a natural resource, it should be properly managed. This means that successful water management is useful in creating a foundation for the achievement of various Sustainable Development Goals. The sustainable development goals have been put in place to ensure availability and sustainable management of water and ensuring sanitation for all. Despite all the measures that have been put in place, water is still becoming a pressing societal and geopolitical issues (M Ongaratto et al., 2018). There are certain regions where water crisis has become a matter of national concern. If the world continues with the "Business as usual" mentality then the world will miss water-related SDGS by a very wide margin. This means that by 2035, about 40% of the world's population will be living in seriously water-stressed areas. This means that the ability of ecosystems to provide fresh water supplies will become highly compromised.

Figure 9.1: Water is a valuable resource.

Source: By José Manuel Suárez, CC BY 2.0, https://commons.wikimedia.org/w/index.php?curid=4602887

Globally, about 60% of freshwater comes from river basins that cross national borders. When dealing with issues of uncertain environment and climatic conditions, governments have to be robust when setting trans-boundary water agreements. The demographic and social changes will raise the global population to 9.7 billion by 2050 and twice the number of people living in urban areas. When not careful, different conceptualizations of water can lead to conflict. There is a common perception that water is a human right and common public and environmental good. This perception has been opposed by the view of water as a commodity that needs to be priced so as to ensure sustainable and efficient use. Peaceful and effective integrated water resources management can be achieved when nations, provinces and communities align water perspectives. This allows sustainable use of water as a resource.

To effectively manage water, governments will need to tackle neglected issues such as water wastage in the current system currently estimated to be up by 30%, unethical practices, corruption in water sectors of many countries, poor accountability and common institutional dysfunction. There a number of reports that highlight the looming water crisis across the world. Water crisis can be described in some various topics including water scarcity and insecurity, sanitation and health, water-related disasters, ecosystem degradation, unsustainable development and water infrastructure deterioration and destruction (Medeiros et al., 2012. There are various bodies including agencies, civil societies and governments that have been involved to deal with issues related to the environment. These groups have made it clear that radical new approaches to water are required to reverse the negative effects on water availability. This can only be achieved by facing the crises in a cohesive and intelligent way. This will ensure that water is able to continue supporting life, biodiversity and development for the young generation and the future generations.

9.2 WATER RISK

When discussing water scarcity, topics related to water risks are not far off. Water risk is defined as the possibility of an entity experiencing water-related challenges such as flooding, drought, infrastructure decay, water stress and water scarcity. When relating water risk to water scarcity, it has been found to be inversely related to water security. This means that a decline in water security causes an increase in water risk. Water risk is often at times described as multidimensional and complex. It covers risks from natural disasters such as drought and flooding that can lead to worsening hunger and infrastructure failure. Realization of these risks results in water scarcity among other problems. There are potential economic effects of water risks that are very significant (Kumar, 2022). Certain industries are very dependent on water. They include the food and beverage, semiconductor and industries, utilities, oil and gas and agricultural industries. Any occurrence of water risk threatens the survival of these industries. For instance, agriculture makes use of about 69% of global fresh water. This makes the agriculture industry vulnerable to water stress.

9.2.1 Water Scarcity and Insecurity

With regards to the global crisis of water as a natural resource is the issue of water insecurity ad scarcity. People may think that there is no challenge with water scarcity as there is a notion that water is plentiful as it covers about 70% of the planet. This notion is not true as only about 2.5%of Earth's waters are fresh. This makes fresh water a limited resource. As a limited resource, it is needed to meet and support about 9.7 billion people in 2050. It is estimated that by 2050, about 3.9 billion or about 49% of the Earth's population will be living in severely water-stressed river basins (Matta & Uniyal, 2017). It is important to note that it's not the population that is pressuring water resources. There is significant evidence to show that there has been excessive utilization of water as a resource.

Figure 9 2: Water scarcity is an issue affecting several states.

Source: By Kaustuvraj - Own work, CC BY-SA 4.0, https://commons.wikimedia.org/w/index.php?curid=97172500

In the 20th century, it is said that the global population tripled but water usage increased to about six-fold. It is expected that between now and 2050, demand for water will increase by 400% from manufacturing and about 130% from household use. With the decline in water availability, it is expected that competition for water as a limited resource will increase. About 60% of surface fresh water comes from internationally shared river basins and about 592 trans-boundary aquifers. For this reason, continuous coordination and cooperation between nations is crucial in ensuring water has been made available for economic, human and environmental needs. There are a number of international water agreements that have been signed over time to allow countries to cooperatively manage growing resource pressure so as to reduce conflicts over water. High levels of water insecurity can lead to drought.

Currently, in some parts of the world, there are large numbers of people who are affected by drought compared to any other natural disaster. For example, in 2016 about 411 million people were affected by disasters that account for 94% of those affected by drought. Drought is also considered a costly disaster that has significant impacts on agriculture. With regards to losses, droughts account for about US$6-8 billion of losses in agriculture in the United States in a year. In China, the occurrence of drought results in an annual grain production loss of more than 27 million tons in the last two decades (Lund & McCullough, 2008).

In China, from the 1950s to the beginning of this century, the average crop area in a year suffering from drought has increased from 11.6 million Hectares to 25.1 million Hectares. This is an increase of about 116%. There are various buses of freshwater across the world. If water was to be secured for irrigation, the potential global welfare gain for reduced risk would be US$94 billion in the year 2010. It was also noted that there has been enhanced water security useful in stabilizing food crop production and prices. For a water-secure scenario, there is a decrease from 83%-38% in the likelihood of global wheat production falling below 650 million tons in a year. When tackling topics pertaining to water scarcity, it is important to note that there are two types of water scarcity namely economic water and physical water scarcity. They are all different in various ways. Physical water scarcity is used to describe a scenario where there is not enough water to meet the demands including the amount of water needed to ensure effective functionality of an ecosystem. This is well demonstrated in arid areas such as North Africa and West Asia known to suffer from physical water scarcity.

Figure 9.3: Human activities have contributed to increased levels of water scarcity.

Source: https://www.shutterstock.com/image-vector/harmful-effects-human-activity-on-planet-1824375041

Economic water scarcity is brought about the lack of investment in technology or infrastructure to draw water from aquifers, rivers and other water sources. It can also be caused by the insufficient human capacity to satisfy the demand for freshwater. This problem is mostly experienced in Sub-Saharan Africa. The temporal and geographic mismatch between the availability and demand for freshwater is the essence of global water scarcity. At the global level on a yearly basis, there are sufficient amounts of water available to meet the demand for freshwater. However, there are temporal and spatial variations in water demand and availability that are large in nature. This results in physical water scarcity in certain parts of the world during certain times of the year. The rapid increase in world populations, changing consumption patterns, expansion of irrigated agriculture and improved living standards are considered the main driving forces for the global increase in demand for freshwater (Leipe et al., 2014). Insufficiencies in the water supply can be caused by climate change, wasteful use of water, increased water pollution and deforestation. There are variations in water scarcity over time attributed to natural hydrological variability. The variation may be a function of prevailing economic policy, management and planning approaches. Water scarcity may lead to the intensification of most forms of economic development. Most causes of water scarcity can be mitigated or avoided.

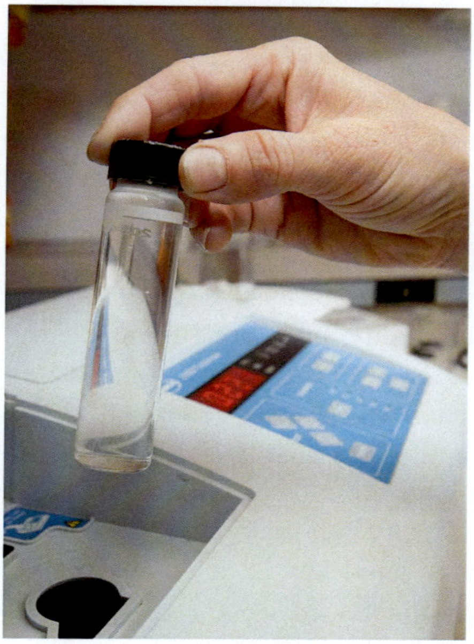

Figure 9.4: Various data are utilized when making assessments on the level of water quality.

Source: By U.S. Department of AgricultureLance Cheung/Photographer/USDA photo by Lance Cheung - 20150409-RD-LSC-0080, Public Domain, https://commons.wikimedia.org/w/index.php?curid=64976496

To tackle issues of water security, there are a number of assessments that have been tone to incorporate information on soil moisture, globalization, virtual water trade, environmental flow requirements and water quality. Dealing with the problem also involves collaborations between social science, aquatic ecosystems science, water quality and hydrological science communities. Among the parameters used in measuring water scarcity is water stress. This is categorized under Sustainable Development Goals. About two-thirds of the global population is said to live in conditions of severe water scarcity for at least one month of the year. It is expected that about half a billion people in the world face severe water scarcity throughout the year while half of the world's largest cities are said to encounter scarcity. Among the option used in reducing water scarcity include cooperation between countries, expanding sources of usable water through wastewater desalination or reuse, water conservation through the prevention of water pollution, supply and demand side management and Virtual water trade.

Water security can be considered as a policy goal used in describing the fundamental societal goal of water policy and water management. This allows the harnessing of productive potent water while evaluating the destructive impacts and possibly eliminating them. There are several productive contributions of water to ecosystems and humans. It ensures their well-being, development and livelihood. On the other hand, water can also be considered to have destructive effects on ecosystems and societies when there is too much water causing floods or when there is too little water causing water scarcity and drought and finally when there is poor quality water (Larsen & Alp, 2015). For this reason, the widely accepted definition of water scarcity is that water scarcity is the reliable availability of an acceptable quantity and quality of water for health, production and livelihood, coupled with acceptable levels of water-related risks. Issues pertaining to water scarcity can be improved by properly managing water resources. This is considered key to achieving growth, poverty reduction and sustainable development.

Figure 9.5: Natural disasters such as flooding are among the risks associated with water.

Source: By johndal - https://www.flickr.com/photos/johndal/2835286257/in/photolist-5jxzPP-GMYPh-dxZ6YV-99EMz9-dgJGFN-dxSC3t-2jsbwH-5jBTQ5-5jxB8X-gF2qp-GMYQf-5jBUBC-5jxBgX-99ENnh-dgJP17-dgJMcb-

9id1Bv-7XTr9i-7PFpsk-5jxz1x-7XWG8j-ampKGU-ampJwY-dgJE6G-5jBSSY-9N5KiH-9ig5N1-dgJQK1-5puhdW-ampMuj-dgJN3h-qXd4uv-4YfCmX-rQv3W4-dy5AQm-7PFpAT-9c7vP2-7cwNtp-5jxARv-9BAN4A-8ChMYf-7PFp-tx-99P3vb-6zszRC-k7x8Dw-7PJAVs-9bWQFh-Ure6Cq-iYJaX4-aHKMLg/, CC BY-SA 2.0, https://commons.wikimedia.org/w/index.php?curid=62293788

There are three main factors used when determining a society's ability to sustain water security. These include socio-economic environment, hydrologic environment and changes in the future environment also known as climate change. Ensuring water security is very crucial when meeting the Sustainable Development Goals. This is attributed to the fact that SDGs cannot be met without access to adequate and safe water.

9.3.1 Physical Water Scarcity

This type of water scarcity is exhibited in areas where there is not enough water to meet all demands. This includes the water needed to ensure the effective functioning of an ecosystem. Physical water scarcity is mostly experienced in arid regions. Climate has been greatly influenced by human beings more so by the activities they undertake. Most of these activities have led to increased levels of water scarcity in areas that were previously hard to come by. This kind of water scarcity also occurs in areas where water seems to be abundant but the resources are over-committed. This mostly occurs when there is overdevelopment of hydraulic infrastructure often used in irrigation or energy generation.

Figure 9.6: Arid regions mostly experience physical water scarcity.

Source: https://www.shutterstock.com/search/arid

Physical water scarcity can be indicated by severe environmental degradation such as water pollution, the decline in groundwater and water allocation that tends to favor some groups over others. This kind of water scarcity is usually as a result of inadequate natural water resources required when supplying a region's demand and economic water scarcity brought about by poor management of the readily available yet sufficient water resources. Poor management of the available water resources has been found to be more often the cause of countries or regions experiencing water scarcity. This is according to the United Nations Development Programme. This is explained by the fact that most regions or countries have enough water to meet household, environmental, agricultural and industrial needs (Knight et al., 2015). They however do not have the means to provide it in an accessible manner. For this reason, about a fifth of the world's population lives in areas affected by physical water scarcity.

In densely populated arid areas, water is physically scarce. This includes areas such as North Africa and Central and West Asia. It is estimated that there are availabilities of less than 1000 cubic meters per Capita each year. From the studies done, it was established that there are more than 1.2 billion people living in areas of physical water scarcity. This kind of water scarcity relates to water available to be used in food production rather than the amount available for drinking. This means that water available for drinking is in small amounts.

9.3.2 Economic Water Scarcity

In areas such as the Western Sindh Province in Pakistan, people collect clean water for drinking from a tap stand. Poor investment in technology or infrastructure to draw water from aquifers, rivers and other water sources is the leading cause of economic water scarcity. It is also caused by the insufficient human capacity to satisfy or meet the demand for water. About a quarter of the world's population has been affected by economic water scarcity. This kind of scarcity includes a lack of infrastructure causing people without reliable access to water to have to move over long distances to fetch water. In most cases, such kind of water is contaminated by rivers for agricultural and domestic uses such as irrigation. This kind of water scarcity is mostly experienced in a significant part of Sub-Saharan Africa. To deal with this problem, it may require governments to invest in water retention and irrigation infrastructure. This approach would be very useful in developing countries that have a heavy dependence on low-yielding agriculture.

Ensuring community members get access to water that is adequate for consumption would greatly benefit the people. This type of scarcity can only be overcome when there are sufficient resources. This means that the process needs more than just new infrastructure. It will make use of socio-political and socio-economic types of interventions that seek to address socio-inequality and poverty. Most developing countries are faced with the challenge of a lack of funds. This means that much effort should be placed into the planning phase. So much emphasis has been placed on improving water sources for domestic purposes and drinking (Knott et al., 2002). This is because large amounts of water are usually used in cleaning, livestock tending, laundry and bathing while small amounts are used in drinking and cooking alone. Such kind of observations can prove that putting too much emphasis on drinking water need addresses an insignificant part of the problem of water resources. This limits solutions that can be used to solve the problem.

9.4 WATER-RELATED DISASTERS

Water-related infrastructure is very important when handling water crisis issues. This makes it vital for governments to protect investments made in water-related infrastructure from shocks and stresses. For instance, it was estimated that by 2030 around half of the bank's water sector portfolio that was around US$8.8 billion was committed and around US$11.3 billion in the pipeline would be at high risk to medium risk of exposures to effects of climate change. Currently, a large number of deaths has been caused by hydrologic hazards. These hazards also lead to displacement and injuries. About 90% of all disasters experienced on the earth are water-related. In the last two decades, the most frequent global natural disaster is floods. In 2016, about 50%of all recorded events were related to flooding. By 2050, the total value of assets expected to be at risk from flooding is predicted to be US$45 trillion. This will be a drastic increase from 2010 by over 340%. There was an 87% increase in population between 1970 and 2010 (Kazanci et al., 2004). This means that the population increased from 3.7 billion to 6.9 billion. In the same period, the annual average population that was exposed to floods increased from 33.3 to 70.4; which is a 112% increase. It is expected that by 2050, there will be an increase in flood-prone lands, deforestation, rising sea, climate change and loss of wetlands.

Figure 9.7: Hurricanes are among the various kinds of water-related disasters.

Source: By J.T. Blatty / FEMA - https://www.fema.gov/media-library/assets/images/142033, Public Domain, https://commons.wikimedia.org/w/index.php?curid=64272928

These levels may increase the number of people who are vulnerable to flood disasters to numbers higher than 2 billion. In 2015, the UN was prompted to release warnings about urban flash floods. This was a response after hundreds died in Guatemala, southern France and the USA. Studies of these events showed that urbanization, intense rainfall and change in climate have led to the frequent occurrence of such disasters in the last two decades. Mitigation of water-related disasters can be done in water-related ecosystems. In disaster-related services, every hectare of mangrove and coastal marsh is worth up to US$15,161per year. The coastal wetlands have been useful as they aided the government and other services avoid more than 625 million in damages from Hurricane Sandy in 2012. Growing of coral reefs is very useful as they act as wave barriers. They also act as an example of their effectiveness in risk education. For this reason, about 1 million dollars are used annually in restoring reefs at the Folkestone Marine Park located on the west coast of Barbados (Jeong et al., 2008). They could be useful in lowering annual storm losses. However, there is a decline in water-related ecosystems at a global level despite the risk reduction benefits. In parts of Asia and America, about half of all coastal mangrove ecosystems have been destroyed or degraded.

9.5 WATER, SANITATION AND HEALTH CRISIS

So much effort has been placed to help increase the supply of drinking water to a larger number of people and households in a year. In 2015, it was said that about 663 million people did not get any access to improved drinking water. For large numbers of people, the provision of improved water does not necessarily mean that water is safe, affordable, accessible or reliable with equity. A good example is Bangladesh where around. 45 million people drink water containing arsenic concentrations greater than concentrations listed in WHO standards. However, significant achievements have been achieved with regard to hygiene and sanitation. There is still room for improvement as about 2.4 billion people are not able to get access to sanitation facilities. There is great importance for all populations to get equal access to sanitation and hygiene. In a group of 19 people, it is said that about seven of them do not get access to improved sanitation facilities while nine out of ten people are still practicing open defecation and others live-in rural areas.

Figure 9.8: Issues of water, sanitation and health crisis are mostly experienced in developing countries.

Source: By Ring-slab latrine in Kalibari community - https://www.flickr.com/photos/gtzecosan/3682625821/in/set-72157625417563576, CC BY 2.0, https://commons.wikimedia.org/w/index.php?curid=36323290

There are various reasons why sanitation and hygiene is an important topic of discussion with regard to save water availability. The level of sanitation in a given is an indication of the availability of clean water. Clean water is used in ensuring that various areas are clean and well maintained. Sanitation also deals with the handling of garbage. In third world countries, there is poor hygiene and sanitation. This is because water has not been properly managed. Some of the trash finds its way into rivers or lakes used in providing water that will be used for drinking. There are cases where animals feeding on the banks of lakes or rivers defecate in the water body making water contaminated and unsafe for drinking or cooking (Islam & Chowdhury, 2013). This reduces the amount of fresh water available for drinking. In most developing countries, people are forced to drink contaminated water as there is limited availability of fresh water. This explains the large number of citizens suffering from water-borne diseases and other diseases related to sanitation such as typhoid and cholera.

Women and girls are mostly affected by the absence of hygiene and sanitation services. They tend to suffer from health repercussions and personal danger when some or all of these services are not secure or not available. Some diseases associated with poor water and sanitation include diarrhea. Diarrhea accounts for 1 out of 9 child deaths across the world. For this reason, diarrhea is considered the third leading cause of death among children under the age of 5. Some neglected tropical diseases are caused by poor sanitation, hygiene and water. Some tropical disease includes intestinal worms, trachoma and schistosomiasis. These diseases affect more than 1.5 billion people annually. In low and middle-income countries, most of these services may not be available to households, health facilities, schools and workplaces that lack water, sanitation and hygiene. A survey done in 2015 for LMICs showed that about 38% of health facilities do not have an improved water source while 35% did not have access to soap and water to be used for hand washing. It was also noted that about 19% did not have access to improved sanitation.

The water, hygiene and sanitation affects both low-income countries and high-income countertop. For instance, in Canada, there are approximately five thousand homes in the First Nation communities that do not have access to basic water and sewage services. When the First Nation homes are compared to other homes, they were found to be ninety times more likely to be without running water. If there are no changes in rapid action then universal water, hygiene and sanitation as described in SDG will just be met. A report generated by the World Bank shows that the capital investment should increase by about 3 times to ensure that water supply, sanitation and hygiene targets are achieved globally. In another study, it was shown that water, hygiene and sanitation efforts will exceed the current by about 4 times so as to achieve targets laid out in the SDG by 2030.

Figure 9.9: Various initiatives have been taken up to increase the availability of drinking water.

Source: By W.J.Pilsak, CC BY-SA 3.0, https://commons.wikimedia.org/w/index.php?curid=194110

To increase the availability of fresh water, there are various groups that have taken an initiative to help communities get access to clean water more so in developing countries. Some of the actions taken by the groups include improved sanitation facilities through the provision of latrines and toilets that flush into a sewer or safe enclosure. They can also promote good hygiene habits through the education of children and the general public. They can also encourage the public to wash their hands using soap and water so as to reduce diarrhea cases. These groups or organizations can also implement the use of rainwater harvesting systems to allow the collection and storage of rainwater that can be used later on in the provision of water for drinking or recharging underground aquifers (Ishida et al., 2020). They can also find projects that will allow the building of well to allow the extraction of ground water from underground aquifers. They can also adopt measures to ensure children and adults are protected from water-related diseases ensuring that they do not claim the lives of children. They can also develop or promote low-cost solutions such as the use of chlorine tablets or plastic bottles that can be exposed to sunlight to improve water quality. They can also provide homes with home water-treatment systems achieved through the presence of filters, flocculants or solar disinfection ensuring water is safe for drinking. Among the groups involved in making interventions include the WPRO. This organization has committed to making interventions on climate change more so those affecting the environmental and social determinants of health-clean air, sufficient food, safe drinking water and sanitation and secure shelter.

Some of the actions and interventions promoted and supported by WPRO include improving the status of WASH in health-care facilitates, developing of water safety plans, promoting of safety-management rainwater harvesting in schools, healthcare facilities and communities, supporting household water treatment and safe storage initiatives with the aim of improving the quality of drinking water while reducing the transmission of water-borne diseases, supporting the fit for the school and the three-star approaches in school and assisting in monitoring activities.

9.6 WATER INFRASTRUCTURE DETERIORATION AND DESTRUCTION

Within the Millennium Development Goals, larger populations have been counted as served by water supply have been allocated to failed systems of allocation. In Africa, there are about 60,000 new hand pumps constructed to help with the problem of water allocation. With regards to water allocation, the issue is that through houses have taps and are connected to a water supply system, very few houses get access to such water. In most households, they have to look for other means of getting access to such water. A study done in 2006 revealed that about 36% of hand pumps across 21 countries in Sub-Saharan Africa were not functional. This means that the government or organizations involved in the purchasing of such pumps incur losses between US$1.2 and 1.5 billion in their investments. In a year, the total cost to water utilities globally caused by non-revenue water is usually a combination of both commercial and physical losses that amount to about US$141 billion (Hodgson et al., 2001). In Africa and other developing countries, about 45 million cubic meters per day are lost through leakage in water infrastructure. The amount of loss is well able to serve about 200 million people. The gravity of the issue is worsened when water infrastructure is poorly maintained. This problem can be experienced in high and low-income countries.

Figure 9.10: Water pipes are among the infrastructure used to manage water allocation.

Source: https://www.shutterstock.com/search/pipes

A typical case may be one in which the capital investment required to maintain aging water infrastructure in the USA reaches an estimated US$294 in 2940. However, there are issues with funding and if they continue to occur then the needs will be underfunded by using US$144 billion. 2015 saw the revolution of the SDGs. They came into play as there was far less international focus on processes and infrastructure for wastewater treatment, water efficiency and water recycling which contributed to some negative impacts in certain areas. A good example is poorly treated wastewater used in agriculture in low-income countries. Studies show that children of age

8-12 years living in areas probe to wastewater have been seen to have a 75% prevalence rate for issues such as gastroenteritis compared to those living in areas that can access freshwater. This means that there is a 73% higher health cost per child for those living in and using wastewater (Alcocer et al., 2021).

Any occurrence of failures in water systems is often at times considered a governance issue. When supplying water, the water sector has a fragmentation of actors and accountabilities hindering and undermining transparency and economic efficiency creating and opening doors for corruption. Some of the issues often recorded in those areas include poor accountability, corruption, opaque decision-making, unethical practices and institutional dysfunction. Though these cases are identified, they are usually difficult to quantify.

There are cases where damages incurred in water infrastructure are deliberate and result in tremendous local impacts. A good example is the one air strike in December 2015 in Syria that led to a cut in water supplies for 3.5 million people. Some highlights of the situation were that there was quick recovery of pumping though about 1.4 million had continued reduced supply. Water and water infrastructures have been used as military targets in Somalia, Iraq, South Sudan, Sudan, The Republic of Congo, Afghanistan, Libya, Yemen, Yemen, Israel, India, Ukraine and Syria since 2011.

There are various damages to water infrastructure. Some of these structures are involved in the management of wastewater to ensure that it does not tamper with the provision of fresh water. In most cases, municipal councils are involved in the management of wastewater. Most of these systems have been developed for the effective treatment and evacuation of effluents and wastewater. Over the years, there are several improvements in the designs of sewage systems. However, there is microbial induced corrosion in the structures that still presents a challenge for the responsible authorities and engineers throughout the world. The term microbial induced corrosion is used to refer to a situation when sulfates in biological waste are converted into sulfides through biological processes under certain anaerobic conditions below the waste-water line in the sewers. Above the waste-water line, hydrogen sulfide gas is released. These gases are reactive later on resulting in aggressive and corrosive condensates with water. These condensates include sulfurous and sulfuric acid.

Figure 9.11: Sulfur is a reactive gas that forms corrosive condensates with water.

Source: By Ivar Leidus - Own work, CC BY-SA 4.0, https://commons.wikimedia.org/w/index.php?curid=101699872

There are various ways in which these infrastructures are well maintained to allow proper management of wastes. There are different companies offering repair services for such infrastructures. They work on contracts with government offices or organizations to deal with such problems. Most of these companies are able to pre-bag a range of gunite products that fits the needs of the government or organizations. For instance, if there are broken pipes or corroded pipes, there are methods used by the company to deal with such issues (Helm, 2017). Some companies have the necessary tools to crack repair solutions for structural and active water leaking cracks. They also make use of epoxy coating systems to allow the protection of interior walls of concrete tanks. They offer a variety of products that allows them to control the corrosion of reinforced concrete.

Some of these measures can be used to handle infrastructure used in managing the supply of freshwater in various homes. Some of the water supply systems have been automated in that they allow automatic detection of leaking pipes. This will send a notification to those handling such systems allowing them to fit the problem before significant losses are experienced. Companies handling water supply can also manage water infrastructure by ensuring that durable materials are selected when establishing water supply systems. They also replace parts or systems that have experienced any kind of damage. Governments should also ensure that they heavily invest in long lasting and durable water infrastructure.

9.7 UNSUSTAINABLE DEVELOPMENT

One of the challenges that have contributed to the water crisis globally is unsustainable development. With unsustainable development, it means that there is a variation in the effectiveness of water management between countries. This means that a rapid scale-up in effort and resources is required for most countries to achieve the sustainable development goals. It also affects their ability to support other water-related or water-impacted SDGs. In 2016, a study was done that shows that when governments take a longer time to act, it becomes hard to deliver their promises by 2030. It further states that every 3 years of inaction is an indication that the amount of effort required to succeed increases exponentially. Among the SDGs, there is SDG6 which covers water goals highlighting the fundamental importance of water to life and livelihood. The success of this SDG will underpin progress in other SDGs such as urban progress, universal education and human health (Hassan et al., 2018). Among the importance of water is that it helps improve poverty allocation and water resource management impacts where almost all aspects of economic activity such as transport, energy production, industry and food production and security.

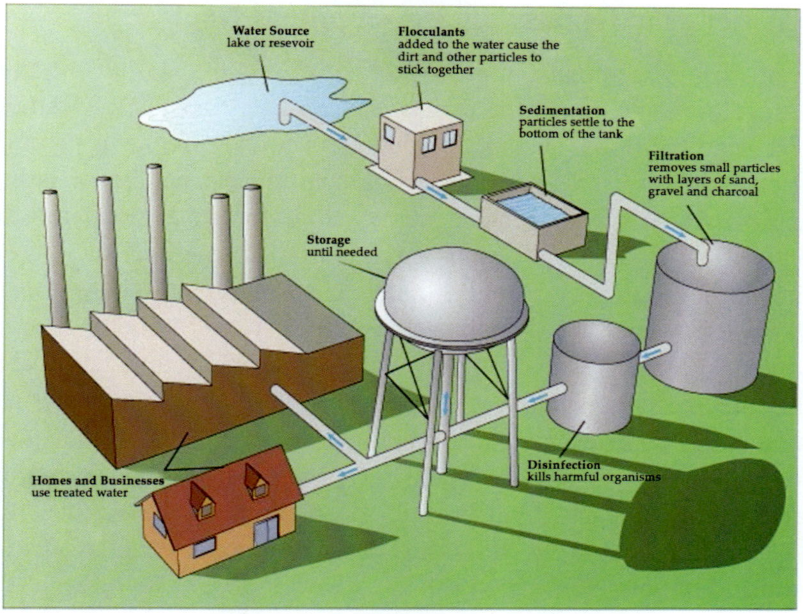

Figure 9.12: Water management is among the goals of SDGs.

Source: By CK-12 Foundation - File:High_School_Engineering.pdf, page 73, CC BY-SA 3.0, https://commons.wikimedia.org/w/index.php?curid=16024233

Water as a resource has been downgraded by human activities. In a day, it is estimated that about 2 million tons of human water are disposed of in water courses. This contributed to the 15-18 billion m3 of fresh water resources contaminated by fossil fuel production every year. This contamination is linked to the foods sector that contributes 40-54% to the production of organic water pollutants in both low and high- income countries respectively. It is worth noting that there is severe pathogenic pollution that affects all rivers and about a seventh of all rivers are affected by severe organic pollution. In one-tenth of all river stretches, there is severe to moderate salinity pollution in Asia, Africa and Latin America (Gothwal, 2021). In order to meet the SDG6 of a state, its government has to ensure that they fully consider the full cost of water and water services provided. Attaining these goals goes beyond simply ticking off sustainability indicators to true sustainability in the water sector.

9.7.1 Water as a Sustainable Development Goal

Water is a key determinant in all aspects of environmental, economic and social development. It was the central focus of any post-2014 framework for food security, global sustainable development, resilience to natural and human-induced disasters and poverty eradication. This can be explained by the fact that water as resource cuts across all sectors and boundaries and is very vulnerable as it is affected by a number of externalities such as social changes, technological changes, climate change, mobile global population, changing lifestyles and consumption patterns and finally economic development. Leaders play a vital role in the setting and actualization of these goals. Some of the indispensable elements useful in building a water-secure world include safe drinking water, water quality, waste water management, development of water resources and the protection of aquatic biological resources, sanitation and hygiene and safe drinking water. Water-related capacity development at the individual bad institutional level is essential in realizing and implementing the Post-2015 Development Agenda.

An overview of water as a sustainable development goal is that there has been a global increase in water usage. Therefore the development and management of water as a resource on earth and the services they provide is unsustainable. Currently, water use has gone up twice the rate of population increase in the last century. This means that there are regions facing increased rates of water shortage. From the years 1990 and 2010, is said that more than 2 billion people have access to basic drinking water while about 780 million people do not get access to such resources. A large population is usually unable to get access to safe and sustainable sanitation. About 1.7 billion people live in river basins where the level of water usage exceeds the rate at which it is recharged. This leads to the desiccation of rivers and the depletion of groundwater. With the increasing rates of population increase and development in urban areas and rural areas, it is expected that demand for water will increase by 55% by 2050. About two-thirds of the world's population could be living in water-stressed countries by 2025 when there is continuous consumption of water (García-Girón et al., 2020). Similarly, it is expected that degradation and climate change of ecosystems as a result of human activities are expected to further exacerbate extreme events such as droughts and floods. Such trends are expected to increase stress in current difficult situations. This also increases the risk of conflict over water. Eradicating property, increased resilience to natural and human-induced disasters can be sustainably achieved by enhancing water productivity. It is also necessary for the government to integrate appropriate infrastructure while integrating an appropriate approach to water resource management and improving water governance systems at all levels. The government also needs to ensure ecosystems are able to support sustainable water management. Some of these ecosystems include aquatic biological resources that may be entirely dependent on water. Water has both good and bad effects on economic growth. Some of the positive impacts include good economic growth, reduction of poverty and provision of power.

More pressure is placed on freshwater resources through water abstraction to allow irrigation and food production. About 44% of freshwater is diverted towards agricultural water withdrawal in OECD countries, 90% in the least developed countries and about 74% in the BRICs. Currently, the high population has led to the global growth in food demand by 60% by 2050. This means larger amounts of water are needed in irrigation while other regions suffer from water scarcity.

9.7.2 Key Challenges

Some of the challenges faced in the sustainable development of water are proper management and lack of funds. Currently, the MDG drinking water targets have been achieved. However, the MDG framework did not full address the broader water agenda that includes issues of water quality, waste water management, living aquatic resources and inter-sectoral competition. Studies show that in about 139 countries, there has been widespread adoption of integrated approaches to water management worldwide. The other challenge is that most of these approaches are usually difficult to apply on the ground. There is uneven progress with regard to improved water governance and efficient water use. This is noted across regions and countries (Gagan et al., 2018). They are constrained by limited implementation capacity and stakeholder participation and suboptimal institutional arrangements. This brings about significant challenges with regards to irrigation, investment in freshwater ecosystem services and rainwater harvesting in areas of concern accompanied by direct impacts on food security and food production.

There are regions where the productive potential of water resources is still untapped as a result of unavailable focused investment more so in Sub-Saharan Africa. Disasters related to water can lead to losses of up to 14% of GDP as a result of a lack of storage and regulation capability. There is a great need for transparent and effective governance mechanisms to allocate water among competing demands. For the world to be truly water-secure, there is a need for cross-sectoral water cooperation at the local, national, regional and global levels. This also needs process engagement with concerned stakeholders.

9.7.3 Measures to Be Undertaken

The unsustainable development of water can be improved by gaining knowledge, innovation, implementation and research. All of these should be geared towards more productive and sustainable use of water more so for energy and food needed to meet the world's future fuel and food needs. All the data used in water management should be updated on a regular basis. It should encompass knowledge of the current status, trends and prospects of water resources and their use are prerequisites for making informed decision-making. Various elements should be incorporated. They include a combination of technical solutions, improved cross-sectoral approaches and political commitment to sustainability so as to meet the competing needs of multiple users. Governments should also provide wiser waste-water management offers to allow enhanced livelihoods through the provision of a safer environment, job creation and general well-being of citizens.

There is a great need for improved global freshwater quality. This can be achieved by addressing water pollution and proper utilization of waste water. It is projected that about 80%of wastewater from human settlements and industrial sources across the world is directly discharged in an untreated form into water bodies. This has detrimental effects on human health and the environment (Eaton & Kardos, 2015). Pollution reduction goes a long way in improving the quality of the world's water resources that requires pollution reduction through the treating and recycling of contaminated water and protecting the ability of the ecosystem in regulating water quality.

9.8 ECOSYSTEM DEGRADATION

The healthy functioning of ecosystems ensures the continuous supply of fresh water. For this to be achieved, it is very important that scientists, research, governments and people recognize the water cycle as a biophysical process vital in achieving sustainable water management while securing the ecosystem used by human beings. There are various water-related services provided by tropical forests that include the regulation of water flows, water purification and treatment and prevention of erosion. They collectively account for a value of up to US$7,235 per hectare in a year. This is more than 44% of the total value of forests thereby exceeding values of carbon storage, recreation and tourism services, timber and food when combined.

Between the years 1997 and 2011, the total worth of ecosystem services lost due to changes in land use ranged between US$4.3 to US$20.2 trillion in a year. More than US$75 billion in goods and services are provided by the freshwater ecosystem for people every year. Ecosystems also sustain disproportionately large numbers of species including a quarter of all known vertebrates.

The challenge however is the fact that most wetlands are facing increased threats to their nature as a result of a host of problems. Analysis of wetlands shows that since 1900, about 64% of the world's wetlands have disappeared. The current level of degradation has been valued at US$29 trillion in lost ecosystem services annually. There are estimates that state that the population of freshwater species has reduced by 76% between 1979 and 2010. About a third of the world's amphibians are at risk of extinction in certain regions while in other areas native freshwater fish species are at risk of extinction by 50%. Wetlands can be referred to as carbon sinks. 3% of the Earth's land's surface is covered by peatlands-lands with pear at the surface-cover. They store about nearly double the carbon and all the world's forest combined when kept wet (Degefu et al., 2014).

It is reported that there has been a general loss of 15% or peatlands. This translates to a contribution of 5% of the world's anthropogenic carbon dioxide emissions. About 45% or half of the peatlands in the Nordic and Baltic States have been drained. In a year, about 80 megatons of carbon dioxide are emitted annually. This is about 25% of the total carbon dioxide emitted by various countries.

Solving water crisis problems requires combined efforts of the general public, the government and various agencies.

9.9 CLIMATE CHANGE AND WATER RISK

Climate change affects the global waters in unique ways. In different areas, higher water temperatures will trigger eutrophication and too much algal development, which will lessen drinking water quality. The quality of fresh water sources may likewise be undermined by high silt amounts or nutrient inputs because of outrageous storm conditions.

Variations in precipitation and overflow timing, combined with higher temperatures because of climate change, may prompt reduced underground water quality. Underground water quality can be preserved or maintained through a blend of watershed management strategies, to diminish pollutant spillover and replenish groundwater sources.

CHAPTER 10

AQUATIC ECOSYSTEMS CONSERVATION AND MANAGEMENT

CONTENTS

10.1 INTRODUCTION

An ecosystem, also known as an ecological system, is made up of a biotope, also known as a habitat, and the creatures that live there. Ecosystems are critical because of the role they play in preserving the overall health of the environment. A biocenosis is a community of organisms developed as a result of the interplay of several environmental factors. Because of the group's unique composition, the ecosystem in which it functions, known as a biotope, is critical to its existence. A biocenosis occurs in a biotope, also known as a "living environment," which contains all of the biotic and abiotic factors that are unique to the habitat in question. The term "habitat" refers to the living environment of a single species or a group of species coexisting (Dyck, 2007). There are crucial distinctions to be noted between habitats and niches, which might be functional or abiotic. An important goal of an ecosystem-based approach is to guarantee that both human and ecosystem needs are met at every level. Here are a few instances, in order of importance: Ecosystems must be adaptive and versatile to persist when faced with fresh scientific, societal, and economic challenges.

Figure 10.1: Marine Ecosystem.

Source: https://media.istockphoto.com/photos/tropical-fish-and-turtle-picture-id539675480?k=20&m=539675480&s=612x612&w=0&h=s20_BSu2Dyy_kmVjqvhQmrE-mde-n0wuw2YjrvnuZoo=

10.2 APPROACH TO WATER RESOURCE MANAGEMENT THAT IS INTEGRATED

According to the Global Water Partnership, an integrated approach to water and land management seeks to increase societal and economic well-being while also conserving the health of vital ecosystems. When it comes to IWRM implementation, GWP prioritizes the following: Management tools, enabling conditions, and a legislative framework According to the International Network of Basin Organizations, basins should be managed at the hydrographic unit level, which includes surface water catchment regions, groundwater aquifers, and basin-based integrated surface and groundwater management (Domisch et al., 2015). To accomplish this, the INBO General Assemblies in Morelia and Fortaleza placed a heavy focus on the need for basin-level agreements, policies, programs, funding, and monitoring throughout their respective sessions.

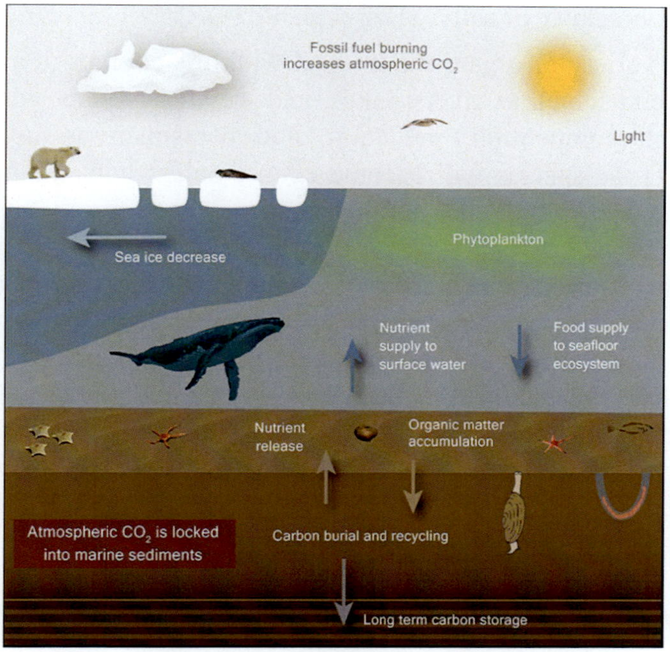

Figure 10.2: Aquatic Ecosystem Pictures Image of a bay with aquatic life.

Source: By Johan C. Faust, Christian März and Sian F. Henley - [1] doi:10.3389/frym.2019.00136, CC BY-SA 4.0, https://commons.wikimedia.org/w/index.php?curid=98736606

10.3 THE IMPORTANCE OF AQUATIC ECOSYSTEMS AND THE FUNCTIONS THEY PLAY

10.3.1 Several Ecosystems Are In Continual Interaction with One Another

Water is an important part of aquatic habitats. Even though aquatic ecosystems cover just a small percentage of the Earth's surface, their existence is essential for the activities and obligations they perform. Consider the following three enormously large categories: rivers, lakes, and wetlands. Rivers, lakes, and wetlands are all dependent on aquatic ecosystems for living. All these settings share a requirement for aquatic ecosystems. This standard defines "natural or artificial" water, "whether flowing or static, fresh, brackish, or salty," as "regions of marshes," "fens," and "bogs," and "water," "whether natural or created, permanent or temporary, with water that is flowing or static, fresh, brackish, or salty."(Dar et al., 2020).

A wide variety of organisms interact with one another in a variety of ways in habitats such as rivers, lakes, and wetland regions, for example. Cohabitation, competition, predation, and parasitism are only a few examples. These species will become extinct unless humans intervene to rescue them. They get the life and nutrients they need from the water, soil, and air in their habitat. This suggests that the makeup of these populations is controlled by a variety of characteristics, including water current and depth, temperature, oxygenation, and chemical content of the water, as well as the catchment region's beds and landforms. In contrast, the presence of living creatures alters the chemical makeup of water, namely the mineral content and the number of dissolved gases present (Cardoso-Silva et al., 2018). As a result, organisms interact with one another and with their environment. It is required to accomplish a complex act of balancing that involves many energy and matter exchanges.

Key Points:

We may be able to learn more about the current status and diversity of aquatic ecosystems by evaluating their structural aspects. Water science needs may be divided into four categories: biological, hydromorphological, chemical and hydrological. The health of aquatic ecosystems in each of these regions may be assessed using several approaches. You can use either qualitative or quantitative variables. When taken together, these measures show the "ecological status" of aquatic environment globally and across time, such as the introduction of the Water Framework Directive in 2000. They are a collection of several metrics for monitoring (Benito et al., 2020).

The majority of the disturbances that occur in aquatic ecosystems are caused by human activities. The status of the aquatic environment may be judged by applying the criteria to an examination of the pressure spectrum produced by these activities when they utilize aquatic habitats. In most cases, these stresses influence biological components. The DPSIR framework, which stands for Driving Forces, Pressures, and State Impact Responses, is one possible foundation for an inquiry.

10.4 INTEGRATED INLAND AQUATIC ECOSYSTEMS MANAGEMENT

Integrated ecological management implies that the environmental issues are seen from a comprehensive perspective considering the biological system as entities and taking into account the wide range of solutions including all potential blends of suggested solutions. The integration implies that every one of the accessible devices and potential outcomes are considered and that every one of the environmental issues are considered. The experience acquired from environmental management across the past 40 years has plainly shown that it is significant not to consider solutions of single issues, rather to consider possibly all serious issues related with a biological system and assess all the necessary management strategies.

Ecosystem Based Management (EBM)

This is one of the most comprehensive ways to protecting aquatic biological systems as it perceives the need to safeguard the whole aquatic environment rather than individual species. Following decades of contamination, habitat degradation and overfishing, presently environmental change and ocean acidification damage ocean health in remarkable ways. Environmental Management represents the ongoing status, patterns, and impacts of the environment, natural disturbances and human impacts on aquatic biological systems. It showcases how to coordinate different management apparatuses and models in an exceptional, multidisciplinary way to deal with environmental management.

10.5 THE FUNCTIONS OF THE NATURAL WORLD

The features of aquatic ecosystems can be used to characterize and/or measure the underlying functioning of such ecosystems. A subset of ecological functions is the relationship between biophysical properties, the existence of biodiversity, and the functioning of ecosystems. The advantages of a healthy aquatic ecosystem are classified into numerous categories, including the following: Ecosystems are responsible for managing the formation of living matter, nonrenewable resources such as water, and mineral compounds, allowing them to maintain the stability of natural processes and the flow of resources. As a result, they are also beneficial in the process of removing dangerous trash that has been changed (Abubakar & Abubakar, 2013). Organizational functions must first be executed for a system to create its own self-organization rules. They are fascinated by the relationship that happens between a system's physical structure and its biological structure.

10.5.1 When Ecosystems Are In Excellent Shape, We Benefit

In a scenario when space is restricted, an aquatic ecosystem is crucial because it can provide a wide range of advantages and services in a little area. Individual gains from ecosystem processes are referred to by scientists as "ecosystem services," which is a catch-all word for all of these advantages. Given the findings of several studies demonstrating the importance of these ecosystems to long-term economic activity, proper water management should be a top priority for ensuring the ecosystems' sustainability. Every ecosystem is based on the activity of numerous species that perform different services. There are several manifestations of ecosystem services that may be discovered. Ecosystem benefits are classified into three types: regulatory, cultural, and provisioning. Management of ecosystem processes leads to regulatory benefits, whereas cultural and provisioning benefits are intangible and do not contribute directly to the economy. The nitrogen cycle, the water cycle, and soil formation are only a few of the processes that help maintain life.

The temporal and spatial expressions of ecosystem services can take many different shapes. Local, national, and even worldwide arrests can be made. The water cycle is a process that occurs throughout the biosphere, and the time it takes to complete one cycle may be measured in billions of years (Bassi et al., 2014).

The pollution produced by people and where it originates in Aquatic environments and the biological services they provide is critical to human survival. This usage may not affect aquatic life at all, or it may have serious consequences known as "dysfunctions," which may be purposeful or inadvertent. The intensity with which it is used decides whether or not it has an impact.

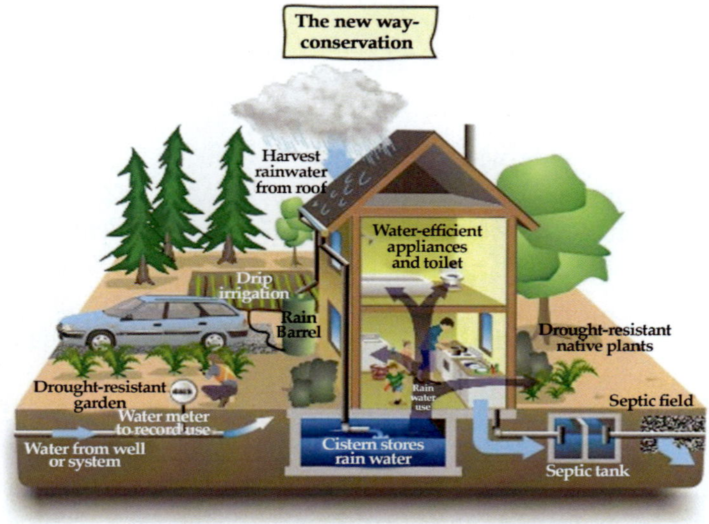

Figure 10.3: Water Conservation.

Source:https://image.slidesharecdn.com/conservation-141123105526-conversion-gate01/95/water-conservation-11-1024.jpg?cb=1416740169

10.5.2 The Biome Has Seen an Increase in the Number of Alien Species

The various aquatic species that dwell in their distinct settings are linked by a food chain and a production pyramid. The loss of one or more species, or the introduction of new species that are not native to the area, might upset this equilibrium. Both the diversity of living things and human health are jeopardized as a result of the introduction of non-native invasive species. All of these plants, as well as others like water primrose and other species like Japanese knotweed, may be found in the wild, in addition to those that have been purposely bred with the idea of being used in landscaping, and vice versa (Balarabe, 2006). Because of the dominance of these invasive plants, native species are being forced out of their habitats, and natural aquatic ecology is being suffocated, putting fish and other aquatic organisms at risk.

Invasive alien plants have the potential to damage humans. Individuals who are sensitive to plant pollen, for example, are more prone to have rhinitis, conjunctivitis, asthma, hives, and eczema. Invasive alien species may harm the productivity of agriculture, livestock, and fisheries, as well as the tourist and recreation sectors.

10.5.3 Hydromorphology is the Study of the Mechanics That Regulate Ecosystems

Human activities and water use may produce a wide range of hydromorphological pressures, which can have an impact on rivers and aquatic ecosystems. This study considers all morphological changes that occur in water bodies, such as changes in the coastline, riparian and littoral zones, water level, and flow. Such forces may be seen in the construction of dams and embankments, the formation of channels, and the manipulation of natural water levels. The principal contributions to the strains on the hydromorphology in the watershed basin are a diverse range of human activities. These include hydropower generation and flood protection system installation, as well as navigation and agricultural techniques enhancement. The hydromorphology of a region is subject to change as a result of a wide range of human activities.

When a river is dredged or channeled with stones or cement, its natural equilibrium is disrupted, affecting the aquatic environment's ecosystem. As a result, the structure of the biotope changes, affecting the organisms that live there. As a result, the species that reside there suffer as well. When there are hydraulic concerns, a river must typically be dredged. If flooding develops as a result of variables like siltation and bed blockage, for example, a dredging contractor may be required. Deforestation, ditch removal, improperly channeled drainage, and other similar activities are common causes of increased sediment erosion at the ground level. Dredging a river may be essential if sediment makes passage difficult or impossible. Mechanical work on the watercourse's bed may be necessary if the bed is colonized by plants growing along its banks, aquatic plants are cultivated there, or the bed is obstructed by a variety of deposits (Barbosa et al., 2012). Dredging rivers for the aim of gathering gravel or sand can harm the natural environment. Human activities like farming and forestry, as well as the construction of blind canals, have a direct influence on aquatic ecosystems.

10.5.4 Hydrology is Made Of the Disciplines of Extraction and Management

A wide range of human activities makes use of water derived from differing aquatic environments. Water can be collected directly from the aquatic environment, or specialized infrastructure can be used to alter the water cycle that happens during the hydrological year (a time period of 12 months for which precipitation totals are measured). Excessive water extraction has caused a significant reduction in the amount of accessible water, putting the ecological balance in peril. When a watercourse is diverted away from the main riverbed, the components of an ecosystem known as hydromorphology and biocenosis might be affected. Changing the annual water curve can affect both the dynamics of water flow and the transport of silt. If a sill or dam is built in a river, it may have a detrimental influence on the aquatic ecosystem's living community.

Some examples of human impacts on hydrology that may be linked back to abstractions: Water from the Aral Sea has been depleted for use on cotton plantations, causing the lake to dry up. The loss of wetlands globally illustrates the extent of this dysfunction, which may have an impact on regions much more confined than the Aral Sea itself (Wu et al., 2021). Watercourse diversions and other hydromorphological alterations that affect the water curve also fall under this category since they disrupt the natural flows of low- and high-flow channels. This is because these sorts of changes affect the water curve.

10.5.5 Chemical and Physicochemical Reactions

The presence of organic contamination is caused by a wide range of human activities. In addition to untreated residential or industrial effluent discharges, agriculture, particularly animal husbandry and land-based manure/sludge spreading can be a significant source of pollution. Agriculture can also contribute to pollution. This organic waste is harmful to aquatic ecology because it reduces oxygen levels, kills fish, and adversely disrupts the ecosystem. Organic pollution may offer a significant quantity of nutrients to aquatic habitats, depending on their composition. When too much fertilizer is applied, both leaching and soil erosion is increased. When there is an overabundance of nutrients in the environment, a process known as eutrophication occurs.

The rapid growth of aquatic plants causes a decrease in the amount of dissolved oxygen in the environment, which leads to hypoxia. Nitrogen and phosphate molecules that sink to the bottom of ponds may be naturally present owing to plant degradation. Human-caused eutrophication, on the other hand, causes an "algal bloom," which eventually leads to the death of fish and the decomposition of other organisms. Human activities may result in the production of micropollutants. Micropollutants are a type of pollutant that, even at very low concentrations, can have a significant impact on the aquatic environment (Tundisi et al., 2010). Some of the most common types of compounds are those formed by the combination of organic elements, metallic elements, and organometallic elements.

The evolution of these fundamental materials can be traced back to a variety of human activities, including raw material extraction, processing, and consumption. It is also possible that it will occur when a person is nearing the end of their life. These substances' physicochemical and toxic properties have a variety of environmental effects. People may be exposed as a result of their surroundings, the food they eat, or the jobs they hold. Ecosystems are susceptible to influence from a wide range of other factors. The structure of the biocenosis is disrupted when hot water discharges occur regularly.

Figure 10.4: Water conservation animated.

Source:https://media.istockphoto.com/photos/water-conservation-in-the-our-planet-usa-picture-id488368309?k=20&m=488368309&s=612x612&w=0&h=-Ojp2OFIa4Xd0ttfQlTaXGPrcSV6ovr1g5iIAYn6ysU=

10.6 RIVER BASIN PLANNING STRATEGIES

The following steps must be taken to successfully plan a project:

Step 1 (Characterization): To accurately characterize a basin, not only the hydrographic basin but also the basin as a whole must be considered. This includes considering the interactions that occur between the activities that occur there, the water resources, and the aquatic environment. Step 2 (Comprehensive Diagnosis): Collaboration is required throughout the diagnostic phase to ensure that environmental factors are considered right from the start of the operation. When it comes to diagnosing basins, there are numerous methods and technologies available.

The primary goals are determined in the following stage, which is based on the previous two stages (characterization and comprehensive diagnosis). There may be greater goals to be achieved through the maintenance and restoration of aquatic ecosystems. Step 3: The next phase is the development of basin-wide plans and procedures, which must include actions that have an impact on aquatic ecosystems (Togaki & Katano, 2020). Step 4: The end outcome of the process is the action plan or program of measures, which is the culmination of the phases that came before it. Indicators of evaluation and support for the measures should be included in the document, as well as instructions on how the measurements will be supported.

10.7 ACHIEVING THE SUCCESS OF BASIN MANAGEMENT PLAN

An approach that takes into account many aspects at once is required to successfully include aquatic ecosystem management, conservation, and restoration into basin management plans. For a basin management plan to be termed "green," the "ecological" component of the plan must be integrated with all of the plan's aims and issues. Including a chapter on ecosystems in the plan is not sufficient on its own.

Before the basin management plan can properly handle environmental problems, it must first address several difficulties, including the following:

Develop a deeper comprehension of the functioning of ecosystems, including the challenges and opportunities that come with them. For the vision and goals of the basin to be successful, environmental factors need to be taken into account. Find out what the ecological goals are for the basin (Teffera et al., 2019). Each basin region must establish its unique objectives, priorities, and levels of protection. The planning process needs to be bolstered by financial planning, which should include planning for water infrastructure, with ecosystems and natural infrastructure being taken into consideration.

10.8 BARRIERS IN THE PATH OF PROGRESS

River basin management techniques must be mindful of the difficulties in incorporating environmental concerns. To begin, environmental issues should be explicitly stated to the basin's organization. A strategy that is more practical for the environment is preferable to one that is unrealistic and difficult to implement. A basin management plan should encompass ecosystems even if there is a lack of understanding before waiting for further information. A plan's design, complexity, or excessively ambitious ambitions could lead to unanticipated technological issues. An implementation may be impeded if the duties of stakeholders are not clearly stated. Addressing the more unclear future trends that influence ecosystems is crucial. The planning process demands open communication. The information and communication that is provided to the public and stakeholders should be well-prepared, consistent, and accurate.

10.9 MANAGEMENT AND RESTORATION OF AQUATIC ECOSYSTEMS

10.9.1 Restoration Practices

As can be seen, the breadth and complexity of restoration undertakings may vary greatly. These initiatives try to restore the natural state and utility of a river, lake, or wetland to assure long-term and multifunctional utilization. River restoration is a critical component of integrated river basin management for long-term water management. The European Center for River Restoration has created a network of best practices for river restoration across Europe as a consequence of its initiatives. Flood risk management is a critical component of many restoration operations.

10.9.2 Methods of rehabilitation

A restoration effort needs knowledge and experience from a variety of fields. The construction phase comes after the planning and design phases.

Preparation and open communication among all parties are important to the success of a restoration effort. As previously said, one reason is that having clear objectives and tracking achievements makes it easier to discover the expertise required for a project. Another purpose is to enable the project to be examined after completion (Soranno et al., 2010). Aquatic ecosystem restoration involves policymakers, practitioners, scientists, non-governmental organizations, and any affected citizen groups. As a result, project managers can better justify their actions and related costs. Restoration work is often governed by a variety of laws. To ensure the success of a project, a complete list of regulatory constraints must be prepared.

Design work will aid in determining the most efficient and cost-effective restoration options for the situation. This is also an excellent opportunity to develop a monitoring strategy, including objectives, geographic scope, duration, measuring technology, and parties involved. This is a fantastic opportunity. Monitoring is required both before and after the completion of a project. The current status report serves as the foundation for pre-project monitoring. Monitoring allows us to establish if ecological objectives have been fulfilled and restoration operations have been effective. Keeping track of customer comments might lead to the development of new projects.

When designing a restoration project, many variables must be considered. Environmental problems, such as erosion and sediment management, noise control, and so on, are often considered while designing projects. While developing a project, all existing features, such as roads, animals, rivers, and wetlands, should be considered. The preservation of recognized historic artifacts, as well as the subsequent assessment of the landscape, are critical issues. Finally, project management and communication abilities are required. Following project completion, it is critical to undertake a post-project review to establish if the repairs matched the original goals. We should share our accomplishments and failings to the best of our abilities so that others might learn from and build on them.

Risk Management in Flooding Situations Rivers, floodplains, lakes, and wetlands contribute to river management both economically and ecologically. During times of floods, they play an important role in the storage, recharge, and outflow of groundwater (Sigala et al., 2017). Connecting brooks, streams, and rivers to floodplains, ancient meanders, and other natural storage sites, as well as improving the quality and capacity of wetlands, may all help to increase natural storage capacity.

10.9.3 Producing Hydroelectricity via Dam Construction

If dams and weirs are utilized to produce power or for other reasons, they represent a major threat to river systems. Dams blocking sediment transport cause both silt retention upstream and sediment loss downstream. This kind of activity has a substantial effect on ecological continuity because it prevents fish from moving upstream and downstream to locate breeding grounds. Migrating fish may be seriously harmed by hydroelectric turbines. Various environmental conditions, such as changes in flow, lake water quality, and temperature, may all contribute to species extinction.

There are many approaches to decreasing the harmful effect of dams. Fish ladders and channels may aid in the migration of a variety of fish species. Measures to maintain structures may help prevent harmful impacts on the environment and animals (Simões et al., 2011). There are numerous ways to help freshwater ecosystems, including reducing downstream fish mortality by shutting down turbines during downstream migration, flushing out sediment with mechanical dam management procedures, installing screens at inlets and fish-friendly turbines, and implementing an environmental minimum flow. Dam operators may also utilize operational constraints like as artificial floods or sediment flushing systems to preserve wetland habitats downstream, which may lower siltation rates in the dam and reduce sediment material shortages downstream.

10.10 ENVIRONMENTAL AND WILDLIFE PROTECTION

Wetland habitats, for example, support a diverse range of animals, including birds and reptiles. River sediments change riverbeds and help ecological development. Humans have changed the structure and aquatic animal habitat of rivers all over the globe by constructing dams, weirs, and watermills to produce energy, make navigation easier and irrigate crops. River restoration increases biodiversity by restoring ecosystems, habitats, and biological continuity. Physical restoration activities include riverbed re-meandering, riverbank restoration to their original green condition, and re-establishing groundwater supplies by restoring the watercourse to its old bed (Shah et al., 2019). Fish tunnels, on the other hand, are built to enable just particular fish species to circumvent river impediments such as sluices and dams. Dams and weirs are usually removed to restore river dynamics when their impounded length or continuity has been harmed.

Wetland and lake environments are critical for species diversity conservation. These areas are home to birds, mammals, amphibians, reptiles, and plants, and they may act as a resting or feeding stop for migrating birds. Wetlands provide a diverse range of organisms that depend on regular flooding cycles for reproduction. Freshwater fish must spawn in wetlands, and many species need shallow wetlands. When wetland restoration efforts are carried out in combination with other natural ecosystems, ecological connectedness improves.

10.11 ENVIRONMENTAL ENHANCEMENTS

The self-purification process that happens along rivers, for example, may play a role in protecting or increasing water quality. Wetlands assist to improve water quality by absorbing nutrients and filtering out contaminants. Wetland regions are sometimes utilized to treat wastewater. Wetland restoration not only helps to preserve these functions but also lays the framework for the development of a healthy ecosystem.

10.11.1 Land Development

Agricultural and forestry activities, for example, may affect river health and the ecosystems that sustain them. Using fertilizer near streams, lakes, or wetlands may result in water pollution due to runoff and projection. Rivers and their tributaries are essential because they offer habitat for a range of plant and animal species. As a consequence, waterways are polluted. Pollution in waterways such as lakes and wetlands may lead to the reduction of biodiversity. Agricultural activities, by transferring exposed soil from the watershed to the water body, may contribute to sedimentation issues. Water plants' capacity to absorb sunlight is reduced, and these sediments cover fish spawning areas. In both cases, buffer strips may act as a physical barrier, preventing toxins and sediments from entering aquatic environments. Forestry may also be done without harming the ecology or morphology of rivers and wetlands. There are plausible technological methods for reducing the effect of equipment or vehicles crossing forest streams through culverts, arches, or bridges.

Nature offers essential services comparable to the vast majority of conventional infrastructure. Wetlands and lakes retain water, as do upland forests and aquifers, rivers, floodplains, and wetlands, while mangroves, coral reefs, and barrier islands protect coasts from storms and flooding. Nature is part of any country's and economy's infrastructure portfolio, which is defined as "the stock of facilities, services, and installations essential to a civilization's operation." As a result, nature functions as "green infrastructure" or "natural infrastructure," as it may supplement or replace the services supplied by conventional infrastructure (Simmatis et al., 2021). The European Commission's definition of green infrastructure includes environmental components that allow communities to make use of green infrastructure's multiple functions.

10.12 INFRASTRUCTURE CREATED BY NATURE

The functioning of natural infrastructure improves the performance of man-made infrastructure. It also provides work opportunities for climate change mitigation and disaster risk reduction. Any project involving dams, storage, irrigation, or drainage must take into account ecosystem services. It may be expensive to exclude nature from planning procedures. To guarantee that the returns on investment for river basin management can be defined, the environmental value must be considered in water investment decisions. Ecosystem services might be priced to provide more precise economic tools for dealing with environmental challenges.

Even if no water is removed from the river or the flow is regulated by infrastructure, changes in land use upstream due to forestry, agriculture, and urbanization may have a significant influence on river flows. Furthermore, the environmental flow may influence groundwater, estuaries, and coastal habitats.

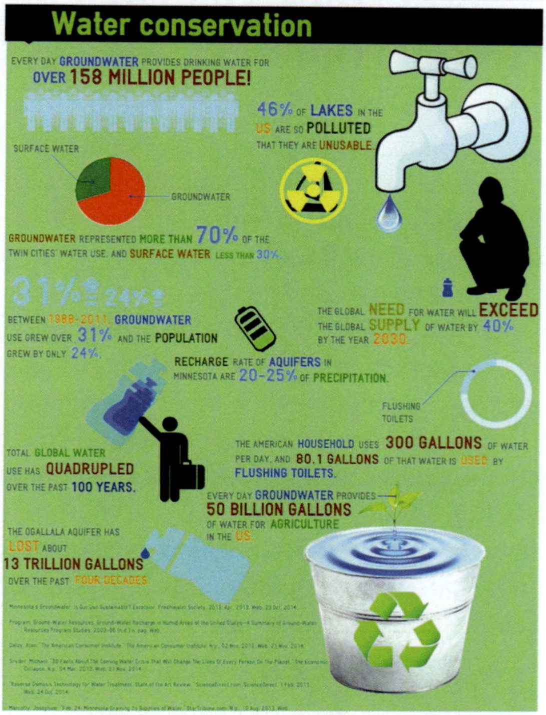

Figure 10.5: A Water Conservation Infographic.

Source: https://i1.wp.com/s3.amazonaws.com/easel.ly/all_easels/631422/WaterConservation/image.jpg?ssl=1

The World Bank has learned the following concerning environmental flow research and implementation. Water resource management must take into consideration environmental flow patterns. Environmental aims must be balanced against social and economic implications. Environmental flow assessments should be performed at each step of the hydrological cycle. During project evaluations, IWRM basin plans establish water allocation guidelines. Continuous monitoring is required to adjust to and enforce flow allocation decisions. Engineering adjustments and reoperations are often needed to deliver the amount of water required for significant ecosystem restoration (Robarts & Zohary, 2018). Economic studies might be utilized to back up the argument for transferring water to downstream areas.

10.13 THE ENVIRONMENT AND THE NEW PARADIGM'S CONSEQUENCES

The International Commission on Large Dams, the International Hydropower Association, the World Bank, and the Economic Community of West African States, to mention a few, have made recommendations for environmental impact studies. According to the World Commission on Dams, human action has separated more than 60% of the world's rivers. In the past, it has been difficult to reduce the environmental implications of large dams. to be specific The construction of fish tunnels to reduce migratory fish obstruction has been ineffective in correcting many of the effects of reservoir expansion on terrestrial ecosystems and biodiversity. Environmental flow requirements are widely used to mitigate the effects of fluctuating stream flow regimes on aquatic, floodplain, and coastal ecosystems. In addition to careful site selection, legislative or political actions may help to minimize or mitigate environmental damage. Implement regional or basin-level mechanisms to guarantee environmental and social standards are satisfied, regardless of funding source. The existing optional template and approach of the European Union Linking Directive should be evaluated. Projects that are currently being funded under the Directive's carbon credit program; Investigate and coordinate the legal interpretation of particular standards or regulations that a donor country has pledged to "respect" or "refer to." Examine the social and environmental consequences of minor actions in the context of the larger picture.

If desired, natural infrastructure may be exploited to generate revenue and invest in grey infrastructure. As a consequence, the river basin comprises a variety of constructed and natural infrastructures that complement one another, yielding the greatest results in terms of cost-effectiveness, risk, and long-term development. The combination of hydropower generation, agricultural water supply, fisheries production, biodiversity conservation, and climate resilience is excellent for achieving a variety of socioeconomic goals. As a consequence, natural infrastructure may serve as a stable foundation for a future economy that is sustainable (Ribas et al., 2019). The study of both green and grey infrastructures has resulted in the development of a new paradigm. Previous research focused on the influence of grey infrastructure on green infrastructure. There are no longer separate inspections for the two types of infrastructure that this new technique examines. Human activity and the natural environment are intertwined in the subbasin under consideration, and they should coexist in the long term.

10.14 MEASURES TO PRESERVE NATURAL WATER ARE IN ADDITION TO THE OVERALL GOALS AND OUTCOMES

10.14.1 Introduction

Devastating floods and droughts have devastated many regions of the world in recent decades, including Europe, Southeast Asia, Australia, and others. As a result, various flood mitigation and preventive projects have been established. According to the British Environment Agency, "engaging with natural processes is becoming increasingly accepted," particularly in "flood and coastal erosion risk management strategy," and these initiatives demonstrate that grey infrastructure solutions alone cannot provide complete protection.

This is why Natural Water Retention Measures (NWRM) are employed alongside other flood mitigation measures such as hard engineering projects. As a result, more and more people are recognizing that NWRM, in particular, has other benefits other than flood control, such as the supply of a variety of Ecosystem Services (Alcocer et al., 2021).

10.14.2 What Does NWRM Stand For?

The acronym "NWRM," stands for "Natural Water Retention Measures". First and foremost, "procedures" or specialized approaches are used to tackle identified concerns and enhance the health of a river basin with NWRM. Secondly, the term "water retention" is an important component of the phrase; it refers to a wide range of strategies aimed at improving water intake by aquifers, soil, aquatic, and water-dependent ecosystems. The co-benefits of this water retention to ecosystems are varied and may be as important as the water retention itself.

"Natural" also refers to a set of techniques used or imitated to meet the objective of water retention by using or imitating natural processes to control the flow and transit of water to avoid extreme occurrences and level out peaks. Based on this data and significant debate, experts in the Water Framework Directive (WFD) Common Implementation Strategy Working Group on the Program of Measures created a European definition of NWRM. Today, several groups and movements have used NWRM for decades, if not centuries. National organizations in nations where natural solutions have already been used to tackle environmental challenges have discovered NWRM-like tactics.

NWRM or equivalent measures are also known as:

In the United States, urban stormwater NWRMs were referred to as "best management practices" and/or "low impact developments" and "sustainable drainage systems" in 2006, according to the DTI Global Watch Mission; in the United Kingdom, the Environment Agency of England and Wales and Newcastle University implemented "runoff attenuation features" on farms in 2011 (Abubakar & Abubakar, 2013).

Agency, the French national organization of water experts and engineers, and Manning-Jones and Southgate have all referred to "Natural Flood Management" as "Catchment Riparian Intervention Measuring."

Figure 10.6: Freshwater fish.

Source:https://www.wallpapers13.com/melanotaenia-boesemani-freshwater-aquarium-fish/

10.15 IMPLEMENTATION OF THE NATURAL WATER RETENTION MEASURES(NWRM)

As previously proven, the notion of NWRM encompasses more than only floods and droughts and the requirement for water storage.

NWRM is extensively used in aquatic ecosystem rehabilitation.

As a consequence, there is a collection of tools that have been utilized in the past but not as part of a comprehensive approach (Reed et al., 2008).

According to the Scottish Environment Protection Agency (SEPA), urban solutions have concentrated on decreasing floods in lower-lying cities and other urban disturbances, whereas agriculture methods have been developed to boost land yield or water management as a crucial component of production.

Land use and NWRM are intricately intertwined.

NWRM deployment needs careful consideration of several factors, including:

When relocating land from places with high water retention to areas with low water retention, keep water retention capacity in mind.

Avoid interfering with the stream's regular hydraulic operation.

Soil sealing in urban areas should be reduced to preserve landscape diversity, organic matter returns to the soil, and green cover while also conserving landscape diversity.

The development of various water retention options, such as permeable pavements and long-term drainage systems, should be encouraged to slow down and profit from the reaction to large rainfall events.

These ideas may be utilized to develop a set of practical NWRM for four mainland uses agriculture, forestry, urban areas, and aquatic ecosystems.

Even if they are employed in several land use spaces, most treatments employ tactics that are particular to the primary land use for which they are designed.

10.16 PASSIVE AND ACTIVE FORMS OF THE NATURAL WATER RETENTION MEASURES (NWRM)

Passive treatments generally take longer to achieve their aims since they rely on natural healing mechanisms. When extensive recovery timeframes clash with management or environmental policy objectives, active interventions are utilized. Restoration attempts are frequently informed by local knowledge and aim to mimic analogous natural structures/features (Barbosa et al., 2012).

There are two sorts of biophysical impacts to evaluate for NWRM purposes:

Water retention's biophysical and metabolic effects are examined.

Legislation in the European Union, in particular, necessitates an integrated approach.

NWRM can assist regulate runoff, minimizing the risk of floods, and increasing water absorption for a variety of reasons.

Watersheds must be planned for and managed in the future. There are various advantages to employing NWRM, and they may be utilized alone or in conjunction with other tactics. In contrast to traditional structural methodologies, NWRM depends on ecosystem processes and services rather than a fully created system .Physical constructions are not restricted in the NWRM, but only the absolute minimum required for ecosystem formation and the provision of associated Ecosystem Services are permitted to be built.

A river basin intervention strategy should aim to make it more ecologically friendly. Rather than focusing on preservation, it should strive to restore any natural environmental functions that have been harmed as a result of grey infrastructure development. If an undisturbed ecosystem or its core functions are restored, climate change, natural catastrophes, and other environmental disturbances may be rendered more robust.

It is possible to do so in structural or non-structural methods.

Ecosystem restoration is a tough management undertaking that requires both inter-and intra-disciplinary efforts (Pinilla & Gabriel, 2016). Environmental restitution is typically part of larger conservation and preservation efforts, both on a policy and legal level.

Sustainable natural resource management and climate change adaptation are now inextricably linked with ecosystem restoration to build a more resilient society. European Union Directives and policies, which were established in the early 1970s to stimulate environmental actions in the EU Member States, must be properly understood by emphasizing several specific aspects that are unique in international law.

First and first, EU rules must be followed, and violators risk fines and harsh penalties.

It's also worth mentioning that in new member states, as well as candidate nations and countries in the European Union, EU Directives are frequently referred to as "blueprints" or "guidelines." Protests and financial incentives are regularly held in response to new laws. Furthermore, EU law is produced through a protracted public consultation process that includes members of the public. Civil society can track non-compliance with the Act by filing complaints with the Commission and Ombudsman concerning insufficient or inappropriate enforcement. Finally, maintaining policy consistency and coordination throughout the EU, as well as internationally and regionally on pertinent topics, is vital.

Despite constraints such as reduced intensity and focus, this customized technique provides a steady driving force (Pinilla, 2010). This implies that the bulk of EU Member States' attempts to restore and conserve ecosystems are influenced by EU laws and policy frameworks, which incorporate a large amount of international environmental law. The global community began to appreciate the value of ecosystems and biodiversity in the 1970s, which led to the implementation of laws that have helped protect natural and semi-wild regions and biodiversity in both the short and long term.

Several international environmental treaties have been signed and ratified in response to growing concern about global warming, including the 1971 Convention on the Conservation of Internationally Important Wetlands, the 1973 Convention on International Trade in Endangered Species of Wild Fauna and Flora, and the 1979 Convention on the Conservation of Migratory Wild Animals in Europe and Natural Habitats in Europe.

Through the idea of "sustainable progress," the UN World Commission on Environment and Development's 1987 "Our Common Future" report attempted to combine economic growth and environmental/ecosystem conservation. Because of the conclusions of this study, a significant new level of environmental awareness and regulation was put in motion at the Earth Summit in Rio de Janeiro in 2007. The adoption of "Agenda 21" and the "Rio Declaration" was one of Rio's most notable achievements (Pérez-Fuentetaja et al., 2006).

Other legally binding treaties, such as the Convention on Biological Diversity, Climate Change, and Desertification Conventions, were also available for signature at the time.

The Rio Summit issued a political declaration emphasizing the need of preserving biodiversity for future generations.

The 1979 Directive on the Conservation of Wild Birds, as codified and amended by Directive 2009/147/EC, the 1992 Directive on the Conservation of Natural Habitats and Wild Fauna and Flora, and the 1998 European Biodiversity Directive are all landmark EU policies and legislation in nature conservation and ecosystem protection/restoration."Biodiversity loss should be halted to achieve this by 2010," EU leaders agreed in June 2001 in Gothenburg, Sweden.

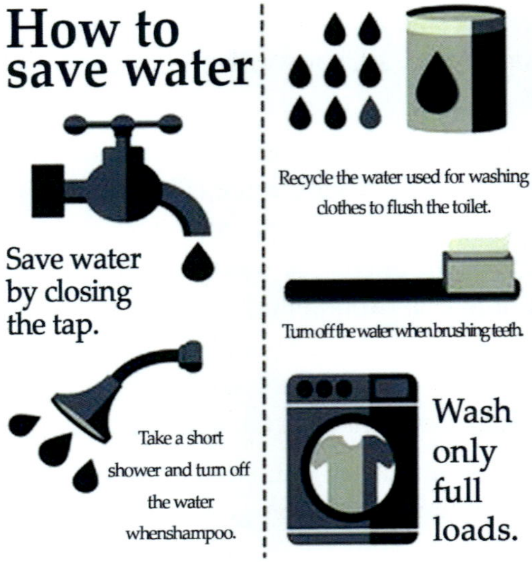

Figure 10.7: Water conservation tips.

Source:https://image.shutterstock.com/image-vector/water-conservation-illustration-conceptual-vector-260nw-234121723.jpg

The 2010 Biodiversity Target was supported by the United Nations General Assembly in 2005 at the World Summit as a result of the Convention on Biological Diversity's sixth Conference of the Parties in 2002, emphasizing the relevance of biodiversity concerns on the global political scene.

In promoting relevant EU policies and legislation, the EU Life - Nature financial instrument was and continues to be an important catalyst and funding source for a large number of pilot/experimental and more traditional nature restoration and rehabilitation projects benefiting European priority habitats, particularly Natural 2000 sites. The European Commission examined this year's EU Biodiversity Action Plan in a joint report from the European Council and Parliament.

Despite significant progress, both the global and EU-specific targets of averting biodiversity loss in the EU by 2010 have not been met (Papa & Mamaril, 2011). A mixed or falling condition was identified for Europe's ecosystem services; nevertheless, the total situation is significantly more concerning, especially as the impact on biodiversity continues to grow.

Land-use changes, pollution, the spread of invasive alien species, and climate change are just a few of the issues we face today. With this year's findings, the European Union and the rest of the world will be better prepared to reach their 2020 targets.

The European Commission published in May 2011 Europe's new strategy to reach the EU's 2020 biodiversity objective, which was declared by EU leaders in March 2010. With this strategy, the EU established a long-term objective of the "EU 2050 vision" and a short-term goal of the "EU 2020 vision." As part of this vision, the European Union's biodiversity and the ecosystem services it offers - its natural capital - must be protected by 2050 for their intrinsic contribution to human well-being and economic development, preventing catastrophic changes induced by biodiversity loss.

EU 2020, more than any other policy effort, aims to increase Europe's commitment to biodiversity protection by 2020, halting the degradation of ecosystem services in the EU Member States by 2020, and restoring them in any manner possible, all while integrating them into the EU's overall development.

As previously stated, the EU's policies and laws on ecosystem conservation and restoration are also linked to climate change and a variety of other global concerns. Since 1992, international climate change conferences have been held in Rio de Janeiro. Since these summits, climate change negotiations have continued, and substantial agreements have been reached, including a second Kyoto Protocol commitment period until 2017. There was also talk of putting adaptive measures in place, such as ecological restoration (Pathak & Mudgal, 2005). When the Green Paper on Climate Change Adaptation was published in 2007, it established an official EU climate change adaptation plan.

According to the Commission's approach, the EU should act early if enough knowledge is obtained; incorporate adaptation into EU foreign relations; fill knowledge gaps, and engage all stakeholders in the development of adaptation plans. Several European Union nations are conducting ecosystem restoration initiatives, the majority of which are sponsored by LIFE, as a result of international agreements and applicable European Union rules.

The growing requirement for coastal flood protection has created a huge opportunity to repair formerly deteriorated coastal wetlands. This is somewhat justifiable in light of the expected loss of intertidal habitats for conservation purposes. Furthermore, coastal engineers have discovered that marshes on the seaward side of sea walls decrease the cost of building and maintain these barriers dramatically.

To create new opportunities for ecosystem restoration, climate change adaptation methods and habitat connectivity have been included in the Climate Sub-Program, Integrated Projects, and Prioritized Action Framework Projects (Owusu-Boateng et al., 2022). Another intriguing characteristic of EU law is EU member states' participation in regional accords, notably those covering the Mediterranean Sea and its coastline region. The 1995 revisions to and renaming of the "Convention to Protect the Mediterranean Sea Against Pollution"(CMPSM) is one such example.

The Barcelona Convention, which was established in 1992, has expanded to include not just pollution reduction but also biodiversity protection in coastal zones. The Barcelona Convention parties created this sustainable development approach in 2005.Both the Oslo and Paris agreements were approved in 1992 by their 14 signatory governments, as well as Switzerland and the European Commission, and the North-East Atlantic Ocean Convention was ratified in 1998 based on these treaties.

The protection and conservation of ecosystems and biodiversity in the maritime region is one of the Convention's four key goals, along with dangerous chemicals, radioactive substances, and eutrophication, and is covered in Annex V. Two concerns are noteworthy because they impact the EU's participation in regional conventions (Ohtaka et al., 2011). First, while it is well acknowledged that UN treaties do not have mandatory enforcement/sanctions provisions, the fact that the EU is a contracting party merges the convention's contents into Community law, making them compulsory for the EU Member States.

Second, in the context of regional or neighborhood policies, the EU funds pilot, demonstration, and other programs, many of which directly or indirectly address ecosystem protection and restoration projects that frequently extend to non-EU neighboring countries, acting as a driving force beyond the EU's borders.

REFERENCES

1. Alcocer, J., Merino-Ibarra, M., Ramírez-Zierold, J. A., Oseguera, L. A., Cortés-Guzmán, D., Castillo-Sandoval, F. S., & Pérez-Ramírez, M. G. (2021). Lake Alchichica limnology. *Lake Alchichica Limnology. The Uniqueness of a Tropical Maar Lake; Springer Nature: Berlin/ Heidelberg, Germany*, 491.
2. Abubakar, M. M., & Abubakar, J. Y. (2013). Some aspects of the limnology of Nguru Lake, northeastern Nigeria. *International Journal of Basic and Applied Sciences*, *2*(2), 140.
3. Barbosa, J. E. D. L., Medeiros, E. S. F., Brasil, J., Cordeiro, R. D. S., Crispim, M. C. B., & Silva, G. H. G. D. (2012). Aquatic systems in semi-arid Brazil: limnology and management. *Acta Limnologica Brasiliensia*, *24*(1), 103-118.
4. Balarabe, M. L. (2006). Limnology in Nigeria: problems and prospects. In *Proceedings of the 11 th World Lakes Conference-- Proceedings* (Vol. 2, pp. 336-339).
5. Bassi, N., Kumar, M. D., Sharma, A., & Pardha-Saradhi, P. (2014). Status of wetlands in India: A review of extent, ecosystem benefits, threats and management strategies. *Journal of Hydrology: Regional Studies*, *2*, 1-19.
6. Benito, X., Vilmi, A., Luethje, M., Carrevedo, M. L., Lindholm, M., & Fritz, S. C. (2020). Spatial and temporal ecological uniqueness of Andean diatom communities are correlated with climate, geodiversity and long-term limnological change. *Frontiers in Ecology and Evolution*, 260.

7. Cardoso-Silva, S., López-Doval, J. C., Moschini-Carlos, V., & Pompêo, M. (2018). Metals and limnological variables in an urban reservoir: compartmentalization and identification of potential impacted areas. *Environmental monitoring and assessment*, *190*(1), 1-13.
8. Dar, S. A., Bhat, S. U., Aneaus, S., & Rashid, I. (2020). A geospatial approach for limnological characterization of Nigeen Lake, Kashmir Himalaya. *Environmental monitoring and assessment*, *192*(2), 1-18.
9. Domisch, S., Jaehnig, S. C., Simaika, J. P., Kuemmerlen, M., & Stoll, S. (2015). Application of species distribution models in stream ecosystems: the challenges of spatial and temporal scale, environmental predictors and species occurrence data. *Fundamental and Applied Limnology*, 45-61.
10. Dyck, M. G. (2007). Community monitoring of environmental change: College-based limnological studies at Crazy Lake (Tasirluk), Nunavut. *Arctic*, 55-61.
11. Degefu, F., Herzig, A., Jirsa, F., & Schagerl, M. (2014). First limnological records of highly threatened tropical high-mountain crater lakes in Ethiopia. *Tropical Conservation Science*, *7*(3), 365-381.
12. Eaton, S. W., & Kardos, L. P. (2015). The limnology of Canandaigua Lake. *Lakes of New York State: Ecology of the Finger Lakes*, *1*, 225.
13. Gagan, M., Kumar, A., Dhingra, G. K., Prashant, S., Laura, G., & Kumar, A. (2018). Limnological assessment of anthropogenic activities of River Henwal. *Journal of Chemical and Pharmaceutical Sciences*, *11*(1).
14. García-Girón, J., Heino, J., Baastrup-Spohr, L., Clayton, J., de Winton, M., Feldmann, T., & Alahuhta, J. (2020). Elements of Lake Macrophyte metacommunity structure: Global variation and community-environment relationships. *Limnology and Oceanography*, *65*(12), 2883-2895.
15. Gothwal, R. (2021). A significant review to Indian limnology. *World Scientific News*, *157*, 105-128.
16. Hassan, F. M., Al-Zubaidi, J. N. A., & Youssef, O. S. (2018). Limnological study of Diyala river, Iraq. *The Iraqi Journal of Agricultural Science*, *49*(3), 452-462.
17. Helm, P. A. (2017). Improving microplastics source apportionment: a role for microplastic morphology and taxonomy?. *Analytical Methods*, *9*(9), 1328-1331.

18. Hodgson, D. A., Vyverman, W., & Sabbe, K. (2001). Limnology and biology of saline lakes in the Rauer Islands, eastern Antarctica. *Antarctic Science*, *13*(3), 255-270.

19. Ishida, T., Uehara, Y., Ikeya, T., Haraguchi, T. F., Asano, S., Ogino, Y., & Okuda, N. (2020). Effects of winter flooding on phosphorus dynamics in rice fields. *Limnology*, *21*(3), 403-413.

20. Islam, M. A., & Chowdhury, A. H. (2013). Limnological status of Trimohini Beel of Rajshahi, Bangladesh. *Journal of the Asiatic Society of Bangladesh, Science*, *39*(2), 173-182.

21. Jeong, K. S., Kim, D. K., Pattnaik, A., Bhatta, K., Bhandari, B., & Joo, G. J. (2008). Patterning limnological characteristics of the Chilika lagoon (India) using a self-organizing map. *Limnology*, *9*(3), 231-242.

22. Kazanci, N., Girgin, S., & Dügel, M. (2004). On the limnology of Salda Lake, a large and deep soda lake in southwestern Turkey: future management proposals. *Aquatic Conservation: Marine and Freshwater Ecosystems*, *14*(2), 151-162.

23. Knott, B., Jasinska, E. J., & Smith, K. D. (2002). Limnology and aquatic fauna of EPP 173, Melaleuca Park, refuge for an outlier population of the black-stripe minnow *Galaxiella nigrostriata* (Galaxiidae), in southwestern Australia. *RECORDS-WESTERN AUSTRALIAN MUSEUM*, *21*(3), 291-298.

24. Knight, S., Cullum, R., & Lizotte, R. (2015). The limnology of a Mississippi River Alluvial Plain oxbow lake following the application of conservation practices. *Journal of Water Resource and Protection*, *7*(09), 709.

25. Kumar, S. (2022). Limnological environment with respect to conservation aspects of Patna Bird Sanctuary Jalesar, Etah (UP). *GSC Advanced Research and Reviews*, *10*(2), 138-145.

26. Larsen, S., & Alp, M. (2015). Ecological thresholds and riparian wetlands: an overview for environmental managers. *Limnology*, *16*(1), 1-9.

27. Leipe, C., Demske, D., Tarasov, P. E., Wünnemann, B., Riedel, F., & Members, H. P. (2014). Potential of pollen and non-pollen palynomorph records from Tso Moriri (Trans-Himalaya, NW India) for reconstructing Holocene limnology and human–environmental interactions. *Quaternary International*, *348*, 113-129.

28. Lund, M. A., & McCullough, C. D. (2008). Limnology and ecology of low sulphate, poorly-buffered, acidic coal pit lakes in Collie, Western Australia. In *Proceedings of the 10th International Mine Water Association (IMWA) Congress. Karlovy Vary, Czech Republic* (pp. 591-594).
29. Matta, G., & Uniyal, D. P. (2017). Assessment of species diversity and impact of pollution on limnological conditions of River Ganga. *International Journal of Water*, *11*(2), 87-102.
30. Medeiros, A. S., Biastoch, R. G., Luszczek, C. E., Wang, X. A., Muir, D. C., & Quinlan, R. (2012). Patterns in the limnology of lakes and ponds across multiple local and regional environmental gradients in the eastern Canadian Arctic. *Inland Waters*, *2*(2), 59-76.
31. M Ongaratto, R., C Loureiro, R., M Restello, R., & U Hepp, L. (2018). Effects of land use and limnology variables on the dissimilarity of common and rare aquatic insects in Atlantic Forest streams. *Revista de Biología Tropical*, *66*(3), 1223-1231.
32. Mustapha, M. K. (2011). Perspectives in the Limnology of Shallow Tropical African Reservoirs in Relation to Their Fish and Fisheries. *Journal of Transdisciplinary Environmental Studies*, *10*(1).
33. Naik, A. A., Wanganeo, A., Ishaq, A., & Bhat, N. A. (2012). Summer Limnology of a high mountain Lake'Kailash Lake'Bhaderwah, Jammu and Kashmir. *International Journal of Environmental Sciences*, *3*(3), 931.
34. Ohtaka, A., Narita, T., Kamiya, T., Katakura, H., Araki, Y., Im, S., ...& Tsukawaki, S. (2011). Composition of aquatic invertebrates associated with macrophytes in Lake Tonle Sap, Cambodia. *Limnology*, *12*(2), 137-144.
35. Owusu-Boateng, G., Ampofo-Yeboah, A., Agyemang, T. K., & Sarpong, K. (2022). Assessment of the limnological characteristics of Lake Bosomtwe in the Ashanti Region of Ghana. *Pollution*, *8*(3), 792-803.
36. Pathak, S. K., & Mudgal, L. K. (2005). Limnology and Biodiversity of fish fauna in Virla Reservoir MP (India). *Environment conservation journal*, *6*(1), 41-45.
37. Papa, R. D. S., & Mamaril Sr, A. C. (2011). History of the biodiversity and limno-ecological studies on Lake Taal with notes on the current state of Philippine limnology. *Philippine Science Letters*, *4*(1), 1-10.

38. Pérez-Fuentetaja, A., Clapsadl, M. D., Einhouse, D., Bowser, P. R., Getchell, R. G., & Lee, W. T. (2006). Influence of limnological conditions on *Clostridium botulinum* type E presence in eastern Lake Erie sediments (Great Lakes, USA). *Hydrobiologia*, *563*(1), 189-200.
39. Pinilla, G. (2010). An index of limnological conditions for urban wetlands of Bogotá city, Colombia. *Ecological Indicators*, *10*(4), 848-856.
40. Pinilla, A., & Gabriel, A. (2016). Experiences in the use of limnological indices in Colombia. *Acta Biologica Colombiana (Online)*, *21*(1, suppl.), 241-248.
41. Reed, J. M., Leng, M. J., Ryan, S., Black, S., Altinsaçli, S., & Griffiths, H. I. (2008). Recent habitat degradation in karstic Lake Uluabat, western Turkey: a coupled limnological–palaeolimnological approach. *Biological Conservation*, *141*(11), 2765-2783.
42. Ribas, L. G. D. S., Padial, A. A., & Bini, L. M. (2019). Advancing impact evaluation in applied limnology. *Acta Limnologica Brasiliensia*, *31*.
43. Robarts, R. D., & Zohary, T. (2018). Limnology and the future of African inland waters. *Inland waters*, *8*(4), 399-412.
44. Simmatis, B., Paterson, A. M., Edwards, B. A., & Smol, J. P. (2021). Linking changes in sedimentary Cladocera assemblages to limnological variables in 67 Sudbury (Ontario, Canada) lakes reflecting various degrees of metal smelter impacts. *Canadian Journal of Fisheries and Aquatic Sciences*, *78*(10), 1455-1469.
45. Shah, J. A., Pandit, A. K., & Shah, G. M. (2019). Physico-chemical limnology of a shallow lake in the floodplains of western Himalaya from last four decades: present status. *Environmental Systems Research*, *8*(1), 1-10.
46. Simões, N. R., Ribeiro, S. M., & Sonoda, S. L. (2011). Diversity and structure of microcrustacean assemblages (Cladocera and Copepoda) and limnological variability in perennial and intermittent pools in a semi-arid region, Bahia, Brazil. *Iheringia. Série Zoologia*, *101*(4), 317-324.
47. Sigala, I., Caballero, M., Correa-Metrio, A., Lozano-García, S., Vázquez, G., Pérez, L., & Zawisza, E. (2017). Basic limnology of 30 continental waterbodies of the Transmexican Volcanic Belt across

climatic and environmental gradients. *Boletín de la Sociedad Geológica Mexicana*, *69*(2), 313-370.

48. Soranno, P. A., Cheruvelil, K. S., Webster, K. E., Bremigan, M. T., Wagner, T., & Stow, C. A. (2010). Using landscape limnology to classify freshwater ecosystems for multi-ecosystem management and conservation. *BioScience*, *60*(6), 440-454.
49. Teffera, F. E., Lemmens, P., Deriemaecker, A., Deckers, J., Bauer, H., Gamo, F. W.,& De Meester, L. (2019). Why are Lake Abaya and Lake Chamo so different? A limnological comparison of two neighboring major Ethiopian Rift Valley lakes. *Hydrobiologia*, *829*(1), 113-124.
50. Togaki, D., & Katano, I. (2020). Detection of freshwater mussels (*Sinanodonta* spp.) in artificial ponds through environmental DNA: a comparison with traditional hand collection methods. *Limnology*, *21*(1), 59-65.
51. Tundisi, J. G., Matsumura-Tundisi, T., Pereira, K. C., Luzia, A. P., Passerini, M. D., Chiba, W. A. C., ... & Sebastien, N. Y. (2010). Cold fronts and reservoir limnology: an integrated approach towards the ecological dynamics of freshwater ecosystems. *Brazilian Journal of Biology*, *70*(3), 815-824.
52. Wu, Q., Sakata, M. K., Wu, D., Yamanaka, H., & Minamoto, T. (2021). Application of environmental DNA metabarcoding in a lake with extensive algal blooms. *Limnology*, *22*(3), 363-370.

INDEX

MIX

Papier | Fördert gute Waldnutzung

FSC® C083411

Kolibri 360 GmbH
Ferdinand-Jühlke-Straße 7
99095 Erfurt, Deutschland
produktsicherheit@kolibri360.de